Microbiology of Cooling Water

James W. McCoy

Supervisor, Refinery Services
Standard Oil Company of California

Chemical Publishing Co.
New York, N.Y.

Chemical Publishing Co., Inc.

Printed in the United States of America

Preface

In a book on water treatment published a few years ago* I included a brief discussion of the subject of microbiology as related to recirculating cooling water systems. The enquiries I have since received and the interest shown in this subject by many readers of that work have encouraged me to expand the original brief treatment to a full-length book addressed to those chemists and engineers who would like some understanding of the microbiology of cooling water

Presumably, people who take the trouble to write books do so either because they think they have something new to say or because they believe they can improve on what has already been published. There are many authoritative textbooks of microbiology and bacteriology, most of which contain a great deal more information on these subjects than is included here. If the reader is interested in the specific topic of the microbiology of cooling water, however, he will find that the literature of this subject is for the most part meager and nonspecific. It consists, in general, of self-serving articles posing as technical papers that contain either no information or else misinformation as to the mode of action of microbicides, proper dosages, effective methods of application, and relative costs. Such publications are full of nonquantitative anecdotal material, dubious conjectures, and uninformative generalizations that are of no use whatever to operators of cooling water systems, nor to anyone else, for that matter.

Microbiology is, thus, a mystery to the average operator of cooling systems, and for this reason he is likely either to discount or to exaggerate the harmful effects of microorganisms. Confusion is rife in this field on account of a strong inclination among the uninitiated to label every instance of foam, slime, or floc formation in cooling water as microbiological. Thus, an enormous amount of money is squandered annually in futile efforts to eliminate chemical slimes and dirt with microbicides. By applying the principles set forth in the following pages, however, it has been possible

* J.W. McCoy, *The Chemical Treatment of Cooling Water* (New York: Chemical Publishing Co., 1974).

to reduce by about 85 percent the cost of microbiological control in a group of cooling systems for which I am responsible. Also, in the years since the first book was published means have been found to predict, in a general way, the quantitative effects of particular microbicides so they can be used more efficiently and economically.

The first three chapters of this work are devoted to a discussion of the kinds of microorganisms that infest recirculating cooling water systems, their effects on these systems, and factors influencing their health and growth.

Chapter IV is a detailed account of biological oxidation processes and of the factors that must be taken into account when using the effluents from these processes for cooling. Reuse of treated effluents is an important method of water conservation that is currently receiving much attention in the petroleum refining and chemical processing industries.

In Chapter V a number of commercial microbicides are considered with respect to their chemical structures, modes of action, toxicities, and other variables affecting their efficiencies when used to control the proliferation of microorganisms in cooling water. Chapter VI is concerned with practical methods of applying these biocides and making quantitative assessments of their effects on microorganisms. Also, a method is described here for exploiting the benefits of the successive application of a flocculant, a surfactant, and a stabilized solution of chlorine dioxide in controlling the growth of microorganisms. Rather recently, chlorine dioxide has attracted some attention as a promising microbicide for use in cooling water. When used with a cationic flocculant and a surfactant, chlorine dioxide in solution has proved to be an extraordinarily efficient and economical toxicant; used alone, however, it is relatively inactive.

The effect of process contaminants on the action of microbicides is a most important consideration in regaining microbiological control of a fouled cooling system. This subject is discussed in Chapter VII, where some examples are given illustrating many of the difficulties that arise with microorganisms in contaminated cooling water. In the final two chapters methods are given for identifying fouling material, staining bacteria, evaluating toxicants, estimating the percentage of microbiological debris in a fouling deposit, and determining the concentration of certain microbicides. These methods have been used over a seven-year period to develop the treatments described in Chapter VI. The procedures for microbiological control described in the following pages are those with which I have had a great deal of personal experience. They work well and I recommend them, but no other person, organization, or company is responsible for them, nor for anything else in this book.

Finally, I must acknowledge my indebtedness to Dr. Charles F. Hinz of the Bio Control Company, for many instructive discussions of the subject of microbiology. Had I not had the good fortune to become associated with him eight years ago, this book could not possibly have been written. My wife, Dolores, has contributed both support and patience to this enterprise, and I thank her most sincerely.

Richmond, California
November 1979

James W. McCoy

Table of Contents

List of Illustrations

I.

Microorganisms in Cooling Water

The water in an open recirculating cooling system is continuously infested with a variety of nuisance microorganisms introduced for the most part in wind-blown dust. Essentially all of these microbes are indigenous to soil; they include bacteria, algae, fungi, and protozoa, in addition to a few multicellular bacterial predators.

Soil is composed of mineral sediments formed by the erosion of rocks, combined with humus, a stabilized complex colloidal organic mixture, which is formed by the biological action of living organisms on vegetation and on each other. Humus and its precursors provide a rich culture medium for microorganisms. Among other things, soil contains amino acids, proteins, polyphenols, methylated polysaccharides, sugars, pentosans, fats, waxes, tannins, and pigments. In addition, more resistant compounds such as cellulose, lignins, humic acid, and polysaccharides are also present, that are slowly degraded to simpler molecules by the metabolism of the resident microbes, particularly *Pseudomonas* and *Flavobacterium.* Fertile soil contains 5–10 percent of organic material with the remainder composed of water and inorganic substances. One gram of rich garden soil may harbor a billion microbes thriving on the nutrient reserve of organic material.

Bacteria, algae, and fungi are diversified into an enormous number of individual species, with many more undoubtedly yet to be discovered. There are, for example, over 90,000 different species of fungi. The bacteria have been subdivided into 3 classes, 12 orders, 2 suborders, 51 families, 10 tribes, and 208 genera. Some 6500 species of green algae alone have been identified.

Originally the classification of microorganisms was based entirely on morphology, i.e., their shape, the arrangement of cells, and presence or absence of capsules, flagella, spores, and the like. Morphology alone soon proved inadequate, however, and now physiological processes are more important in taxonomy. Thus, bacteria are classified as sulfate-reducing, methane-oxidizing, lactic acid-forming, nitrogen-fixing, and numerous others. Several different classification schemes are currently in use, but the most widely accepted is that originally devised by Bergey.[1] For taxonomic purposes bacteria, algae, and fungi are assigned to the Plant Kingdom, while protozoa, rotifers, and such are placed in the Animal Kingdom. In Bergey's system the blue-green algae are included with the bacteria in Phylum Protophyta, but for convenience they are treated here with the other algae in Phylum Thallophyta.

Although many species of microorganisms inhabit soil, relatively few—perhaps a dozen—find the environment of a recirculating cooling water system congenial, and form a significant part of its population. Table 1.1 is limited to those genera most frequently reported; the symbol (*) indicates genera that occur in the largest numbers in cooling water.

TABLE 1.1

Taxonomy of Microorganisms Found in Cooling Water

Plant Kingdom

Phylum I. *Protophyta* (primitive plants)
- **Class I.** *Schizophyceae* (blue-green algae)
- **Class II.** *Schizomycetes* (bacteria)
  - **Order I.** *Pseudomonadales*
    - **Suborder I.** *Rhodobacteriineae*
      - **Family I.** *Thiorhodaceae*
        - **Genus I.** *Sarcina*
    - **Suborder II.** *Pseudomonadineae*
      - **Family IV.** *Pseudomadaceae*
        - **Genus I.** *Pseudomonas**
      - **Family VII.** *Spirillaceae*
        - **Genus II.** *Desulfovibrio**
  - **Order IV.** *Eubacteriales*
      - **Family III.** *Achromobacteraceae*
        - **Genus III.** *Flavobacterium*
      - **Family IV.** *Enterobacteriaceae*
        - **Genus I.** *Escherichia*
        - **Genus II.** *Aerobacter**
      - **Family XIII.** *Bacillaceae*
        - **Genus I.** *Bacillus*
        - **Genus II.** *Clostridium*

Phylum II. *Thallophyta* (undifferentiated plants)
- **Division I.** *Algae*

Subdivision I. *Cyanophyta* (blue-green algae—alternative classification)
- **Order I.** *Chroococcales*
  - **Family I.** *Chroococaceae*
    - **Genus I.** *Anacystis*
- **Order II.** *Nostocales*
  - **Family I.** *Oscillatoriaceae*
    - **Genus I.** *Oscillatoria**

Subdivision V. *Bacillariophyta* (diatoms)
- **Class I.** *Centrobacillariophyceae*
  - **Order I.** *Eupodiscales*
    - **Family I.** *Coscinodiscaceae*
      - **Genus I.** *Melosira**
- **Class II.** *Pennatibacillariophyceae*
  - **Order I.** *Fragilaiales*
    - **Family II.** *Fragilariaceae*
      - **Genus I.** *Fragilaria*
  - **Order IV.** *Naviculales*
    - **Family I.** *Naviculaceae*
      - **Genus I.** *Navicula**

Subdivision X. *Chlorophyta* (green algae)
- **Order II.** *Chlorococcales*
  - **Family I.** *Chlorococcaceae*
    - **Genus I.** *Chlorococcum*
  - **Family IV.** *Chlorellaceae*
    - **Genus I.** *Chlorella**
  - **Family X.** *Scenedesmaceae*
    - **Genus I.** *Scenedesmus*
- **Order IX.** *Ulotrichales*
  - **Family I.** *Ulotrichaceae*
    - **Genus I.** *Ulothrix**

Division III. *Eumycetes* (filamentous fungi and yeasts)
- **Class II.** *Ascomycetes* (sac fungi)
  - **Genus I.** *Chaetomium*
- **Class III.** *Basidiomycetes* (basidia fungi)
  - **Genus I.** *Poria**
  - **Genus II.** *Polyporus*
  - **Genus III.** *Peniophora**
- **Class IV.** *Fungi Imperfecti*
  - **Genus I.** *Phialophora*
  - **Genus II.** *Cytosporella*
  - **Genus III.** *Pestalozzia*

Genus IV. *Sporocybe*
Genus V. *Bisporomyces**
Genus VI. *Acremonium*
Genus VII. *Chalaropsis*

1. BACTERIA

Bacteria are microscopic, single-celled, plant-like organisms that usually reproduce by binary fission, and in general, lack chlorophyll. Some of them secrete a slime that can serve as a binding agent for dirt and chemical precipitates in cooling systems, forming on occasion slimy or sticky coherent masses that may obstruct the circulation of cooling water. Sixty to seventy percent of the bacteria living in recirculating cooling systems belong to two genera: *Pseudomonas* with about fourteen species, and *Aerobacter* with perhaps seven.

1.1 Classification and Structure

Pseudomonas is a sturdy organism, naturally present in soil, able to live at very low oxygen tension, and difficult to kill. Many species also occur in coastal, swamp, and pond waters; they are more-or-less free-swimming in cooling water. Members of the *Pseudomonas* genus obtain their energy either by oxidizing organic compounds or by fermentation, but those that thrive in cooling water live by oxidizing organic substrates to carbon dioxide. Individual cells rarely combine in pairs or chains, but frequently are found in zoogloeal masses; that is, embedded in a diffuse, gelatinous matrix. All of the pseudomonads are nonsporulating, rigid rods, 0.5–1.0 μ in diameter and 3-5 μ long, which are normally motile by polar flagella. They do well in neutral or slightly alkaline cold water. Very few are pathogenic to warm-blooded animals, including man; *Pseudomonas aeruginosa* can cause enteritis, however, which in infants may be fatal.

The movement of bacteria to regions that are hot or cold, light or dark, or chemically attractive is in large degree intentional. They swim by the action of flagella, which are joined to a rod and a set of rings embedded in the cytoplasmic membrane within the cell; the entire assembly is called an organelle. Flagella are composed of flagellin, a single protein without enzymatic activity. They undergo rigid rotation in either direction at the rate of about 50 revolutions per second, using energy produced by oxidative phosphorylation. In bacteria that have several flagella, these gather into a bundle, and rotate in unison.[2] Under the microscope, the motion of bacteria appears to be random. They move in straight paths, but change direction abruptly every few seconds, also stopping briefly on occasion, and sometimes engaging in a tumbling motion. It is now thought that rather than proceeding at random, bacteria react to sensory perceptions, for they approach attractive material purposefully, and retreat hastily from repellants, such as toxic chemicals. This faculty is called chemotaxis.

Species of *Desulfovibrio* are obligately anaerobic, nonsporulating, curved, or spirally twisted rods that reproduce by transverse fission, and thus tend to form chains. This genus grows best at 20-30°C, but can survive temperatures up to 55-60°C. It flourishes where protected from oxygen as, for example, in or under sediment in the basins of cooling towers, but the highly aerated recirculating water makes it difficult for the organism to become established elsewhere in these systems. *Desulfovibrio desulfuricans* can use elemental hydrogen generated at cathodic sites to reduce sulfate to hydrogen sulfide, thus causing corrosion.

The common species of *Aerobacter* are small, nonsporulating, straight rods, certain strains of which contain orange or yellow pigment. Morphologically they are identical to *Pseudomonas* except that they are peritrichously flagellated. *Aerobacter aerogenes* can grow anaerobically by fermenting glucose to butylene glycol, ethanol, and a mixture of organic acids. Most species of *Aerobacter* are encapsulated in a diffuse slime layer. Slime is especially associated with *Pseudomonas*, however, probably because this genus predominates in cooling systems.

A few spore-forming bacteria of the *Bacillus* genus appear now and then. They are nutritionally more demanding, however, than *Pseudomonas* and *Aerobacter*, requiring specific vitamins and amino acids, and most are unable to grow in a medium containing a single organic compound. A few species can ferment sugar, if available, but many are obligate aerobes. Spores form only when normal, healthy vegetative cells experience an adverse change in environment. When this occurs a thick-walled polysaccharide coating surrounds the nucleus within the bacterium, producing an endospore. When favorable conditions are reestablished the polysaccharide envelope is discarded, and the cell again becomes vegetative. Spores can remain viable with negligible respiratory activity for many years. They are killed only with difficulty by chemicals that destroy vegetative cells on contact, because the walls of spores are impenetrable to most chemicals except chlorine. Pasteurization is useful for isolating spore-formers in water; vegetative cells are killed, but the spores themselves can survive for several hours, even in boiling water. The remaining four genera listed in

Table 1.1, if encountered at all, are merely incidental, and of no significance in the practical operation of cooling systems. A brief description may be of interest, however.

Some *Clostridia* ferment cellulose to acetic acid and ethanol, or sugars to butyric acid. *Clostridium nigrificans* is an anaerobic sulfate-reducer native to sewage, soil, and mud; it has not been definitely implicated in anaerobic corrosion. It is a spore-forming thermophile, growing well up to 58°C, and in the wide pH range of five to nine. *Sarcina*, which is sometimes found in small numbers in cooling water, undergoes cell division in three planes forming characteristic regular packets of cells. *Flavobacterium* is rod-shaped, divides transversely, and may form chains, tetrads, or packets of eight cells; the organism produces yellow, orange, or red colonies on nutrient agar. *Escherichia coli* has been reported in a few cooling waters, presumably introduced in the droppings of birds. It is quickly crowded out, however, by the hardier microorganisms that are better suited to the environment.

Crenothrix and *Gallionella* are often mentioned as sources of trouble in cooling systems. These are the so-called iron-bacteria, which oxidize ferrous to ferric ion producing a precipitate of the hydrous oxide mixed with a mucilaginous secretion, the combination being a voluminous brown slime.[(3)] These bacteria sometimes initiate pitting and tuberculation in iron pipes. They are partial to outfalls where water is entering a stream, but are rare in recirculating cooling water as they do not prefer an environment with high oxygen tension.

Another method of classifying bacteria is by their reaction in a differential staining method developed in 1884 by the Danish physician Christian Gram. Briefly, the procedure is as follows: A drop of bacterial culture is transferred to the center of a microscope slide, spread in a thin film, and fixed on the slide by drying. A solution of crystal violet is applied to the smear, followed by a dilute aqueous solution of iodine. The latter, acting as a mordant, forms a deep blue-black compound with crystal violet. Next the slide is rinsed with alcohol or acetone, which leaches the blue-black compound from some bacteria faster than from others. Finally, the slide is drenched with a solution of safranine, then rinsed with water. Organisms that retain the crystal violet (blue) stain are Gram-positive, while those that are decolorized by alcohol or acetone, then subsequently take the safranine (red) stain are Gram-negative. With the exception of *Bacillus* and *Sarcina*, essentially all bacteria encountered in cooling water are Gram-negative. This characteristic Gram staining is confined to bacteria and yeasts; details of the procedure have been given by Shankel.[4]

The living part of a bacterial cell is called the protoplast; it consists of a nucleus and the surrounding cytoplasm. The latter is a colloidal suspension of proteins, lipids, and carbohydrates, containing granules of reserve food, including fats, polysaccharides, and volutin, a complex mixture of polyphosphate, magnesium, lipoproteins, and ribonucleic acid. All of this is enclosed by the ctyoplasmic membrane, a firm structure 60-100 Å thick composed of 75 percent protein, 20-30 percent phospholipids, and 2 percent carbohydrates. A rigid cell wall 10-23 mμ thick surrounds the cytoplasmic membrane, which in a healthy bacterium, is tightly pressed against the cell wall by osmotic pressure. The effective diameter of the pores in the cell wall is thought to be about 1 mμ. According to Salle,[5] the cytoplasmic membrane contributes 10 percent and the cell wall 20 percent of the dry weight of a bacterial cell.

The Gram-positive cell wall is composed mainly of a highly polymerized glycopeptide loosely linked to a teichoic acid; it also contains lipoprotein and lipopolysaccharide complexes. The repeating unit in the polymerized glycopeptide is acetylglucosamine linked to acetylmuramic acid, which in turn is joined to a short polypeptide.[6] In Gram-negative bacteria the cell wall is three-layered, consisting of mucopeptide with layers of lipopolysaccharide and lipoprotein with a large proportion of polysaccharide. As the rigidity of the cell wall depends upon glycopeptide, the wall ruptures in its absence. The cell wall is a rigid semipermeable membrane that plays a significant role in the metabolism of the organism by serving as a physical barrier against the environment, while the cytoplasmic membrane is the osmotic barrier controlling the transfer of metabolites and waste into and out of the cell. The osmotic pressure in Gram-positive bacteria averages 20 atmospheres, in Gram-negative bacteria about 5 atmospheres.

Gram-negative bacteria have long fine hairs called fimbriae, or pili, attached to their cell walls, whose function is a matter of speculation, although they may serve as a means of adhesion to surfaces.[7] These filaments are composed of a protein pilin and seem to originate on the outer surface of the cytoplasmic membrane from where they penetrate through the cell wall. They are helical structures 1-10 micrometers long and 60-70 Å in diameter; they can be observed under an electron microscope after shadowing with tungstic oxide. The number of fimbriae on individual Gram-negative cells varies from none to over two hundred, and although easily removed by mechanical agitation, they regenerate rapidly. It would be expected that these tangles of long hairs surrounding bacterial cells would tend to trap inorganic particles and other cells, and so play a roll in the formation of microbiological fouling in cooling systems.

Surrounding the cell wall is a third membrane, the capsule. This is a modified outer layer of the cell wall, which if diffuse and gelatinous, is called a slime layer. Its composition varies, but in the main it is a secretion of polysaccharide that arises in the cytoplasmic membrane; it may also contain glycoproteins, protein-polysaccharide complexes, or polypeptides. Eighty to ninety percent of the bacteria form a capsule; it serves as a protective device, but it is never an essential part of the cell. Slime is a normal product of metabolism, and is apparently made up of degradation products of the cell wall; profuse slime is evidence of age and low metabolic activity.

1.2 Metabolism and Growth

The metabolism of all living organisms consists of a number of interrelated enzyme-catalyzed chemical reactions wherein energy is released and new cellular material is synthesized. The production of enzymes in a living cell is regulated by deoxyribonucleic acid (DNA), which is composed of phosphate groups linking deoxyribose units, each of which has attached to it a purine or pyrimidine base, most frequently adenine, guanine, thymine, and cytosine.[8] Two of the resulting molecular strands, consisting of phosphate-sugar chains, are combined by hydrogen-bonded pairings of purine and pyrimidine basie groups, adenine with thymine, and guanine with cytosine. This sequence of bases maintains the genetic code responsible, among other things, for the formation of specific enzymes. The form of the intertwined strands is that of a double helix of constant diameter.[9,10] The role of hydrogen bonds is, therefore, vital to the ontogeny and proper functioning of cells,[11] so chemicals that break these bonds are lethal. For this reason, it is desirable when using enzyme poisons to eliminate bacteria, that these chemicals be supplemented by agents that disrupt the hydrogen bonding in nucleic acids, otherwise inactivated enzymes will be immediately replaced by normally functioning ribonucleic acid, and the organism will survive. The structures of enzymes and coenzymes are also stabilized by hydrogen bonding, consequently agents that rupture these bonds attack both the functioning enzymes and the RNA that produces them. These considerations are important in the control of the growth of microorganisms in cooling systems; they are discussed in greater detail later.

Two important sets of biological catalysts are the digestive enzymes, or hydralases, and the respiratory enzymes, the cytochromes (cellular pigments). The latter function as terminal intermediates for the transfer of hydrogen or electrons to cytochrome oxidase. Hydralases of particular interest in cooling water management are those secreted by certain bacteria into their environment to catalyze the hydrolytic decomposition of the substrate. Extracellular enzymes from cellulolytic bacteria and fungi, for example, degrade cooling tower lumber and cellulosic plant residues to simple fatty acids, which in turn serve as food for other microorganisms in the water. When bacteria die they are decomposed by their own enzymes, an economical process called autolysis, the products of which then provide food for surviving organisms.

The cytochromes are porphyrins, i.e., organic compounds of four pyrrole rings bridged with methine groups, and containing an atom of iron firmly bound to all four nitrogen atoms by either covalent or coordination links.

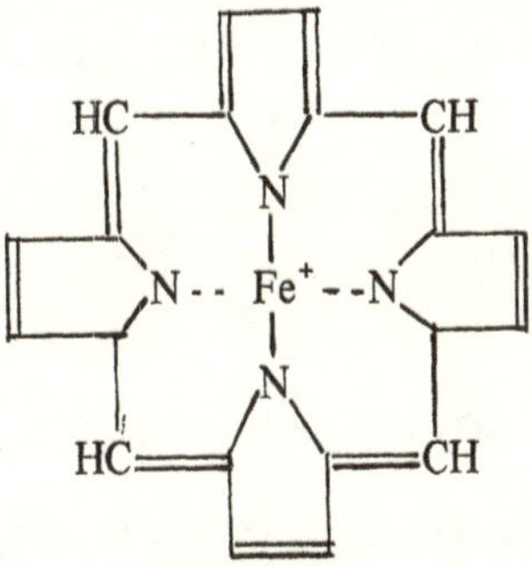

The formula shown is the simplest iron porphyrin; in cytochromes there are numerous substituents on the four pyrrole rings so that the molecular weight of the cytochromes is about 16,000. In aerobic respiration, typical of *Pseudomonas*, *Aerobacter*, and *Flavobacterium*, electrons are transferred from nutritional material to oxygen through the agency of two cytochromes in which the iron is reversibly oxidized and reduced.

$$Fe^{++} = Fe^{+++} + e^{-} \tag{1.1}$$

Iron thus serves as a transfer agent for electrons between the nutritional substrate and oxygen in a series of oxidation-reduction steps, two of which can be formulated as follows:

Cytochrome oxidase reacts with oxygen by transferring electrons from atoms of ferrous iron.

$$4Fe^{++} + O_2 = 4Fe^{+++} + 2O^{--} \tag{1.2}$$

The oxidized enzyme is next reduced by hydrogen atoms liberated by cytochrome hydrogenase.

$$4H + 4Fe^{+++} = 4H^{+} + 4Fe^{++} \tag{1.3}$$

The hydrogen ions then combine with the reduced oxygen to form water.

$$4H^{+} + 2O^{--} = 2H_2O \tag{1.4}$$

These steps, of course, are only formal; the actual process of respiration is much more complicated.[12]

An example of an anaerobically-respiring organism is *Desulfovibrio desulfuricans*, in which the sulfur atom in sulfate, sulfite, or thiosulfate accepts electrons, and is reduced to hydrogen sulfide.[13] Two instances of this process are the oxidation of acetic acid to carbon dioxide, and the oxidation of hydrogen to water, with the concomitant reduction of sulfate to hydrogen sulfide.

$$2OH^{-} + CH_3COOH + SO_4^{--} = 2CO_3^{--} + H_2S + 2H_2O \tag{1.5}$$

$$4H_2 + SO_4^{--} = H_2S + 2H_2O + 2OH^{-} \tag{1.6}$$

Eq. (1.6) represents cathodic depolarization through a dehydrogenase enzyme whereby *Desulfovibrio* reduces sulfate to hydrogen sulfide. The enzymes of anaerobes must be maintained in a reduced state for the organisms to remain viable. Thus, oxygen is lethal, and only 2 ppm of chromate ion inhibits the growth of sulfate-reducers,[14] probably because of the relatively large oxidation potential of the chromate ion.

2. *ALGAE*

Algae inhabit soil in large numbers and, like bacteria, are introduced into open recirculating cooling systems in wind-borne dust. They are not sensitive to pH, and some species can live at 50–60°C, or higher, as well as at normal temperatures. As shown in Table 1.1, the algae most often found growing in cooling towers belong to three subdivisions: *Cyanophyta* (blue-green), *Bacillariophyta* (diatoms), and *Chlorophyta* (green). The majority of these are aerobic, photosynthetic, and reproduce by cell division or vegetative propagation (fragmentation), although spores of some sort have a part in the life-cycles of most species.

2.1 Classification and Structure

The most commonly encountered genera of the blue-green algae are *Oscillatoria*, a filamentous type, and *Anacystis*, a capsulated form that is found in colonies held together in a mucilaginous matrix. *Cyanophyta* prefer a neutral or alkaline environment, so are rare, or absent below pH 6. All of them are enclosed in a gelatinous sheath that corresponds to the capsule in bacteria, and a few secrete toxins that can cause sickness or death in waterfowl and animals. *Mycrocystis*, for example, secretes an alkaloid that adversely affects the liver and central nervous system.[15]

Diatoms that often grow in cooling systems include *Melosira* and *Fragilaria*, both filamentous, and *Navicula*, a unicellular organism. The principal components of algal cell walls are carbohydrates (cellulose, hemicellulose, polysaccharides), and lesser amounts of lipids and proteins. Typically, the cell walls consist of semicrystalline microfibrils of cellulose embedded in an amorphous carbohydrate matrix. Diatoms, however, are surrounded by a wall, or frustule, composed of polymerized orthosilicic acid. The cell wall of diatoms is unusual in that it consists of two sections not firmly connected, which in cell division draw apart to allow a new layer of silicate to form on the exposed surfaces of the daughter cells. The diatoms, particularly *Navicula*, also secrete some gelatinous substances analogous to slime in bacteria.

Ubiquitous genera of the *Chlorophyta* (green algae) are *Chlorella*, *Ulothrix*, *Scenedesmus*, and *Chlorococcum*. The *Chlorophyta* are indifferent to pH, thriving in the wide range of 4.5–9.3. They exhibit two stages of growth: In the first they are free-swimming by means of flagella; in the second, they settle on a solid surface exposed to light, shed their flagella, and become sessile. In time the unicellular types take on a rounded form and enclose themselves in a mucilaginous sheath. *Chlorella* cells occur singly, or in gregarious masses held together by secreted mucilage. *Chlorococcum humicula* is the most pervasive algal species in the world. As it prefers dripping water it is often seen in cooling towers, although it is not the most prevalent species in these structures. When nutrients are deficient, as in a typical cooling system, the filamentous green alga *Ulothrix* predominates. Now and then brown algal mats appear inside cooling towers in places not exposed to direct sunlight, an example of color change in relation to incident light called chromatic adaptation.

The nutrient value of algal protein compares favorably with those of animal and vegetable origin, which led at one time to the study of *Chlorella* as a source of protein to alleviate the world-wide shortage of food. This project was carried far enough to spawn a few recipes for a "*Chlorella* Cook Book".[16]

2.2 Metabolism and Growth

Sunlight, carbon dioxide, and a few minerals are essential for the growth of green algae. These and other photo-autotrophes are provided with chloroplasts, containing chlorophyll and carotenes, that enable them to convert the energy of light quanta from the sun to chemical energy. Chlorophyll is a powerful photo-sensitizer for transforming ordinary atmospheric oxygen to the intensely reactive singlet state. Carotenes, on the other hand, catalyze the conversion of oxygen from the singlet to the triplet multiplicity, and thus protect protein in the cells of green algae and plants from degradation by singlet oxygen. Mutant cells lacking carotene are quickly killed by singlet oxygen in the presence of sunlight. The terms singlet and triplet refer to energy states of an element that are recognized by characteristic spectral lines.[17] The change from singlet to triplet oxygen is brought about when an electron pair is converted to two electrons with parallel spins.

Hill[18] demonstrated that in the presence of a hydrogen acceptor, oxygen could be produced by chloroplasts isolated from green and blue-green algae according to the following equation, which is known as the Hill reaction:

$$A + H_2O \xrightarrow{h\nu} AH_2 + \tfrac{1}{2}O_2 \qquad (1.7)$$

The photosynthesis of carbohydrate can be formulated to show the reduction of carbon dioxide through the agency of some hydrogen donor.

$$CO_2 + 2H_2D \xrightarrow{h\nu} 2D + (CH_2O) + H_2O \qquad (1.8)$$

Calvin[19] has elucidated the mechanism of the entire photosynthetic process, which consists of many intermediate reactions. The net result can be summarized in one equation, however, which when read from left to right represents photosynthesis, and from right to left depicts respiration.

$$CO_2 + H_2O \underset{\text{Respiration}}{\overset{\text{Photosynthesis}}{\rightleftharpoons}} O_2 + (CH_2O) \qquad (1.9)$$

Practically all species of *Chlorella* and *Scenedesmus* can grow in the dark by the oxidative assimilation of glucose or galactose, but very few other substrates are usable by them. In metabolizing glucose these genera produce a mixture of carbon dioxide, and acetic and other organic acids. In general, the diatoms are unable to grow in the dark—that is, they are obligate phototrophs. Also, as their cell walls are composed primarily of silica, the availability of ortho-silicic acid is essential. Other things being equal, the proliferation of diatoms is proportional to the silicic acid available.

A few genera of blue-green algae can grow in an environment that provides only atmospheric nitrogen, carbon dioxide, water, sunlight, and a few essential minerals, notably potassium and magnesium. These are autotrophic organisms that require only inorganic material to synthesize their protoplasm; carbonates or bicarbonates serve equally well as their source of carbon. Many green algae can also fix nitrogen, and some of them, if nitrogen-starved, can assimilate undissociated ammonium hydroxide to satisfy their nitrogen requirement. Oswald[20] has determined the average composition of algal cells; he formulates their synthesis according to Eq. (1.10).

$$1.0NH_3 + 7.62CO_2 + 2.53H_2O \xrightarrow{h\nu} C_{7.62}H_{8.06}O_{2.53}N_{1.0} + 7.62O_2 \tag{1.10}$$

Lewin[21] points out that phosphorus is one of the major nutrient elements required for the normal growth of all algae, principally because phosphorylated compounds are essential intermediates in the reactions of photosynthesis. Only ortho-phosphate can be assimilated, but the organisms convert it to polyphosphate, which is then stored in volutin granules within their cells. In common with most other microorganisms, algae use sulfate as a source of sulfur for the synthesis of such amino acids as methionine, cystine, and cysteine.

3. *FUNGI*

3.1 Classification and Structure

Fungi are simple plants without chlorophyll, and undifferentiated into roots, stems, or leaves, which reproduce by sporulation. They live in soil and are immotile, but their spores are blown about by the winds, in which manner they are introduced into open cooling systems. The osmotic pressure within mycelial fungi is high, consequently they possess a stout cell wall, which is composed of chitin, a polymerization product of acetylglucosamine. According to Albert,[6] the cytoplasmic membrane in fungi is easily penetrated by lipophilic substances, yet few specifically antifungal chemicals have been discovered. A typical higher fungus when mature consists of a mass of cytoplasm enclosed in an elaborate network of multibranched tubes called hyphae. The entire structure is termed a mycelium, which by increase of cytoplasm and extension of the hyphae is capable of indefinite growth.[12] Normally, a single hypha grows at its tip, steadily becoming longer, but when a round colony evolves from a spore, hyphae extend in all directions from the center. The hyphae of certain fungi are highly resistant to lysis and decomposition; this is attributed to the presence of polymeric aromatics in the cell wall.

The fungi of interest to operators of cooling systems are the *Ascomycetes* and the *Fungi Imperfecti*, that produce soft-rot, and the *Basidiomycetes*, which cause white-rot and brown-rot in wooden structures. *Basidiomycetes* is the most prevalent class in cooling towers, specifically the genera *Poria* and *Peniophora.* White-rot is produced by three species; *Poria negrescans*, *Peniophora mollis*, and *Polyporus versicolor.* Brown-rot is caused by *Poria oleraceae* and *Poria monticola.* The genera of *Fungi Imperfecti* most often isolated from redwood cooling towers are *Sporocybe*, *Acremonium*, *Phialophora*, *Cytosporella*, *Bisporomyces*, *Pestalozzia*, *Chalaropsis*, and *Cephalosporium.* The *Ascomycetes* also are associated with soft-rot, but this class is less common in cooling towers, although the genus *Chaetomium* has occasionally been isolated. Many other fungi (and bacteria) are found in wooden structures, especially when decay is present, but they have not been definitely implicated as agents of the deterioration.

Lumber from the California redwood tree *Sequoia sempervirens* is widely used for constructing cooling towers because it is particularly resistant to fungal decay. There is a fungus, *Poria sequoiae*, that infests the heartwood of the living tree producing "brown heart rot," or "pocket rot." Oddly enough the disease is never fatal to the standing tree, and the fungus dies promptly when the tree is converted to lumber.

3.2 Metabolism and Growth

Saprophytic fungi obtain their food osmotically from plant and animal sources by secreting extracellular digestive enzymes into their surroundings. These organisms oxidize sugars and polysaccharides to some forty different organic acids including citric, oxalic, fumaric, and succinic acids.[22]. Cellulolytic fungi use cellulose as a source of carbon, converting it to glucose and cellobiose, thereby destroying wood.

The optimum temperature for growth of a great many fungi is 25–30°C, but most species found in cooling towers prefer 34°C, or even higher.

All of the fungi mentioned here grow optimally near pH 6. *Basidiomycetes*, however, are inhibited by the mildly alkaline environment at pH 7–8. In general, their metabolism is easily disturbed by changes in pH, and by the presence of heavy metals.

4. HIGHER FORMS

The basin of a cooling tower and the microscopic organisms living in its sediment approximate to some degree the ecology of a polluted pond. Dirt blown into a cooling tower is collected by the water raining down through the structure, and much of it accumulates as a sediment on the bottom of the basin. Similarly, during shutdowns water stagnates in various parts of the cooling system, solids settle, an anaerobic environment develops, and holozoic (solid food-eating) protozoa and *Desulfovibrio* proliferate. The most frequent genera are *Amoeba* and *Paramecium*, although small roundworms, the ciliate *Vorticella*, and the rotifer *Philodina* have also been seen. All of these belong to the Animal Kingdom, and are predators of bacteria.

Amoeba is a single-celled organism without fixed form that extends its cytoplasm into finger-like projections with which it engulfs its food, and then absorbs it into the body of the cell. It is native to soil and stagnant water, feeds principally on bacteria, and reproduces by binary fission.

Paramecium is a ciliate protozoan that is virtually covered with short whip-like extensions known as cilia. It also has an oral groove on one side of the cell with rows of cilia that sweep food into a sort of primitive gullet. The surface of the organism contains numerous trichocysts—minute hair-like organs—that secrete a substance, which when wet forms long, sticky threads in the water.

Vorticella is a bell-shaped ciliate with cilia around the open end whose coordinated beating carries food into the body cavity. The narrow closed end of the bell is furnished with a long stalk with which the animal attaches itself to some relatively solid surface. The body waves around in the water as the organism seeks food.

Rotifers are so called because the pattern of the beating cilia around their anterior end has the appearance of wheels spinning. These animals are coenocytes, that is, organisms composed of multinucleate cells. They develop from a single cell by repeated cell division, but when the animal is nearly mature the membranes between many of the cells disappear, and a coenocyte is produced. *Philodina* is a rotifer sometimes seen in cooling tower basins. It is transparent under the microscope so that the transfer of material into and out of the organism, as well as a good deal of streaming and churning within the body, is clearly visible.

None of the higher microorganisms mentioned here is of any significance in day-to-day operations of cooling systems, but as they are now and then encountered this brief description has been included.

References

(1) Buchanan, R. E., and Gibbons, N. E., Eds. 1974. *Bergey's manual of determinative bacteriology.* 8th ed. Baltimore: Williams and Wilkins Company.

(2) Berg, H. C. 1975. Bacterial behaviour. *Nature* 254: 389.

(3) Starkey, R. L. 1945. Transformations of iron by bacteria in water. *J. Am. Water Works Assoc.* 37: 963.

(4) Shankel, D. M. 1968. Elementary techniques in microbiology. *The American Biology Teacher* 30: 461.

(5) Salle, A. J. 1973. *Fundamental principles of bacteriology.* 7th ed. New York: McGraw-Hill.

(6) Albert, A. 1963. Correlations between microbiological morphology and the chemistry of biocides. In *Advances in applied microbiology,* Vol. 5, W. W. Umbreit, ed., New York and London: Academic Press.

(7) Poon, Hung and Day, A. W. 1974. Fimbriae in the fungus *Ustilago violacca. Nature* 250: 648.

(8) Davidson, J. N. 1960. "The Biochemistry of Nucleic Acids", 4th ed., Methuen, London.

(9) Watson, J. D. and Crick, F. H. C. 1953. Molecular structure of nucleic acids. *Nature,* 171:737.

(10) Watson, J. D. and Crick, F. H. C. 1953. Genetical implications of the structure of deoxyribonucleic acid. *Nature* 171: 964.

(11) McClellan, A. L. 1967. The significance of hydrogen bonds in biological structures. *J. Chem. Ed.* 44:547.

(12) Stanier, R. Y., Duodoroff, M. and Adelberg, E. A. 1957. *The microbial world.* Englewood Cliffs: Prentice-Hall.

(13) Lewis, R. F. 1965. Control of sulfate-reducing bacteria. *J. Am. Water Works Assoc.* 57-1011.

(14) Drummond, J. P. M. and Postgate, J. R. 1949. "A Note on the Enumeration of Sulfate-Reducing Bacteria in Polluted Water and on Their Inhibition by Chromate", *J. Appl. Bacteriol.,* 18:307.

(15) Echlin, P. 1966. "The Blue-Green Algae", *Scientific American,* 214, 75 (May, 1966).

(16) Morimura, Y. and Tomiya, N. 1954. Preliminary experiments in the use of *Chlorella* as human food. *Food Technol.,* 8:179.

(17) Pauling, L. and Goudsmit, S. 1930. *The structure of line spectra.* New York: McGraw-Hill.

(18) Hill, R. 1939. Oxygen produced by isolated chloroplasts.

Proc. Roy. Soc. B127:192.

(19) Calvin, M. 1956. The photosynthetic carbon cycle. *J. Am. Chem. Soc.*, 78:1895.

(20) Oswald, W. J. 1961. "Fundamental Factors in Stabilization Pond Design", in *Advances in Biological Waste Treatment*, Pergamon Press, Oxford.

(21) Lewin, R. A., Ed., 1962. *Physiology and biochemistry of algae.* New York and London: Academic Press.

(22) Foster, J. W. 1949. *Chemical activities of fungi.* New York and London: Academic Press.

II.

Effects of Microorganisms in Cooling Systems

At present the majority of recirculating cooling systems draw make-up water from municipal supplies or from wells, in which the number of microorganisms is small, if not zero. Open systems, however, continuously take in dust and wind-borne microbes, which given a certain amount of nourishment, proliferate rapidly, leading to a number of undesirable effects that can interfere with efficient cooling. This problem becomes acute when it is necessary to cool with reclaimed water such as the effluent from a secondary sewage treatment plant. As these streams are relatively rich in organic matter and phosphates, they are nutritious culture media for microorganisms.

1. NUTRITIONAL REQUIREMENTS

A living microorganism is essentially an aqueous system, or gel, composed of about 75 percent water, 10-15 percent protein, and 2 percent minerals, along with lesser amounts of other organic components including carbohydrates, lipids, and nucleic acids. Two mechanisms are used by microorganisms to obtain energy. That most common to bacteria in cooling systems is aerobic respiration in which specific organic compounds are oxidized completely to carbon dioxide and water, with molecular oxygen serving as the electron acceptor. The net reaction using glucose as the source of energy is

$$C_5H_{11}O_5 \cdot CHO + 6O_2 = 6CO_2 + 6H_2O \qquad (2.1)$$

An example of an anaerobically respiring bacterium already mentioned is *Desulfovibrio desulfuricans*, in which the sulfur atom in sulfate, sulfite, or thiosulfate accepts electrons, and is reduced to hydrogen sulfide.[1]

There is a third type of respiration, aerobic fermentation, used by cellulolytic bacteria to obtain energy from cellulose. In this process a carbohydrate in the presence of oxygen is degraded to several different compounds of lower molecular weight instead of being oxidized completely to carbon dioxide. The latter is also evolved, however, and can be used to measure the rate of fermentation.

Cellulose consists of chains of 2000–6000 molecules of D-glucopyranose (β-glucose). (Molecular weight 600,000–2,000,000.)

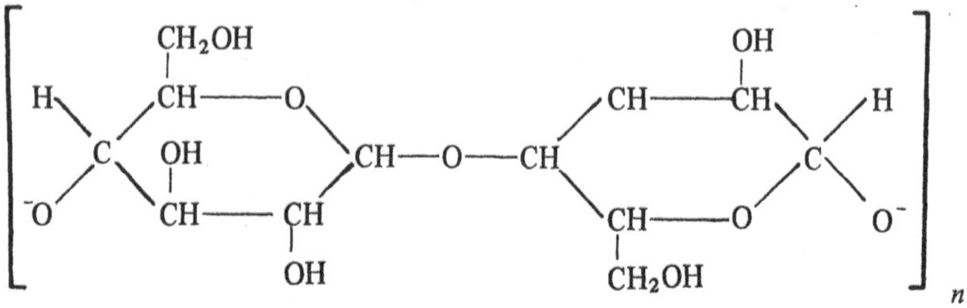

The first step in the fermentation of cellulose yields cellobiose; this is then hydrolyzed to β-glucose, which is further oxidized to mixtures of organic acids, alcohol, carbon dioxide, and hydrogen. Eqs. (1.5), (1.6), and (2.1) represent energy-producing reactions that sustain life, but provide no material for cell synthesis for continued growth. For growth to occur with a single organic compound an organism must be able to use it both as a source of energy in its respiration, and to assimilate it for the synthesis of new cells. An instance of the use of a single organic compound for complete metabolism is the degradation of cellulose within the wooden structural members of cooling towers by the fungus *Poria negrescans.*

Some of the nutrients in clean cooling water come from humus blown into the system in dust. From this source small amounts of carbohydrates, proteins, amino acids, and the like are available to sustain the sturdy types that grow in this unpromising medium. It is likely, however, that a large proportion of the dissolved organic matter from which the bacteria obtain their carbon is provided by algae. Growth factors[2,3] such as thiamine, B_{12}, and other vitamins, as well as certain essential trace elements usually are furnished by the continuous autolysis of organisms that for one reason or another have expired.

The growth of a great variety of algae has been shown to depend upon the amount of phosphate available, which is one of the more important factors required for normal growth.[4] These organisms have other definite requirements that also must be met before they can thrive. Thus, if *Chlorella* is deprived of magnesium, manganese, or potassium there is a drastic slowing of its photosynthetic rate.

2. *BACTERIA*

Within any community of microorganisms certain members predominate on account of their particular aptitude for the environment in which they find themselves. We mentioned in Chapter I that in open recirculating cooling systems exposed to contamination by wind-borne dirt the dominant species of bacteria belong to two genera: *Pseudomonas* and *Aerobacter.* Both of these, along with the less frequently encountered *Flavobacterium*, elaborate profuse gelatinous capsules, which are sticky, and thus can accumulate mud and other debris into masses that clog screens and impede circulation in heat exchangers. Still other bacteria attack cellulose, corrode steel, damage concrete basins, and generate objectionable odors.

2.1 The Formation of Slime

Aerobacter produces a more extensive slime capsule than does *Pseudomonas*, but profuse slime is particularly associated with the latter, which generates material ranging in consistency from that of molasses to rubber. In general, differences in appearance, quantity, or consistency of slime are not inherent in the organisms generating it, but rather are connected with the availability of nutrients. An exception is *Acetobacter xylinum*, which elaborates a copious slime layer of pure cellulose. This species, however, is not indigenous to cooling water; all of the slimes encountered in cooling systems are composed primarily of polysaccharides.

The formation of voluminous capsules is associated in some bacteria with low temperature (4-20°C), but in most they form when the environment is nutritionally deficient. In the process of respiration with limited nutrients one-fourth to one-third of the carbon removed from organic substrates is not assimilated into new protoplasm, but is transformed into extracellular polysaccharides instead in the form of slime. Under some circumstances this can serve as a reservoir of food. A capsule may contain more than one bacterium, but on the average the volume of slime is about four times that of a single bacterium. In addition to serving as a binding agent for accumulations of dirt, slime if unchecked can form sticky films on surfaces where heat is exchanged, plug screens and watertubes, and initiate under-deposit corrosion.

2.2 Degradation of Cellulose

The ability to metabolize cellulose does not satisfactorily define a group of bacteria as there are some 50 species divided among seven genera that secrete extracellular enzymes (cellulases) that hydrolyze cellulose. Cellulolytic bacteria found in cooling water include *Pseudomonas*, *Desulfovibrio*, and less frequently, *Bacillus* and *Clostridium*. These bacteria are seldom implicated as the primary agents in the decay of wooden cooling towers, although they are frequently associated with wood-rotting fungi discussed in Section 4.

2.3 Corrosion and Odors

Desulfovibrio desulfuricans contains an enzyme, hydrogenase, that enables it to use elemental hydrogen generated at cathodic sites to reduce sulfate to hydrogen sulfide. Thus, it acts as a cathodic depolarizing agent in anaerobic electrochemical corrosion, a subject that has been thoroughly reviewed by Costello.[5] Iverson[6] has found *Pseudomonas aeruginosa*, a facultative anaerobe, in tubercles formed in the corrosion of aluminum alloys. This is an example of anaerobic corrosion attributable to the formation of an oxygen concentration cell, a mechanism that is well-established for iron. Cathodic areas surround the tubercle where the surface of the metal is exposed to aerated water, but the metal under the tubercle is anodic because the inside of the tubercle is anaerobic. In addition to the differential concentration cell, there is evidence that the corrosion of some aluminum alloys is caused by the removal of atoms of metal from the alloy structure by extracellular enzymes secreted by certain microorganisms.

In cooling systems electrochemical corrosion occurs most often in fouled areas, and in the majority of instances is attributable to *Desulfovibrio desulfuricans*. This organism is frequently found in the sludge in the basins of cooling towers where, using sulfate as a hydrogen acceptor, it produces hydrogen sulfide and ferrous sulfide, which in combination form a malodorous black slime. In severe cases bubbles of hydrogen sulfide rise through the water in the basin causing objectionable odor in the neighborhood of the cooling tower. In chromate-treated systems chromate is reduced by hydrogen sulfide, precipitating a mixture of colloidal sulfur and hydrous chromic oxide, which sometimes causes severe fouling within the cooling system.

$$3H_2S + 2CrO_4^{--} + 2H_2O = 3S + 2Cr(OH)_3 + 4OH^- \qquad (2.2)$$

The corrosion of iron by sulfate-reducing bacteria is rapid, and unlike ordinary rusting, is not self-limiting. Tubercles formed by *Desulfovibrio* consist of an outer shell of red ferric oxide mixed with black magnetic iron oxide, containing a soft black center of ferrous sulfide, under which there is usually a deep pit. When cleaned out the interior surface of the pit invariably has a characteristic bright silvery appearance.

Starkey[7] has postulated the following steps in the cathodic depolarization theory of anaerobic corrosion by sulfate-reducing bacteria:

$$8H_2O = 8H^+ + 8OH^- \tag{2.3}$$

$$4Fe + 8H^+ = 4Fe^{++} + 8H \tag{2.4}$$

$$SO_4^{--} + 8H = H_2S + 2H_2O + 2OH^- \tag{2.5}$$

$$Fe^{++} + H_2S = FeS + 2H^+ \tag{2.6}$$

$$3Fe^{++} + 6OH^- = 3Fe(OH)_2 \tag{2.7}$$

$$4Fe + SO_4^{--} + 4H_2O = FeS + 3Fe(OH)_2 + 2OH^- \tag{2.8}$$

Eqs. (2.3) and (2.4) represent the anodic dissolution of iron. Eq. (2.5), the essential step, represents cathodic depolarization through a hydrogenase enzyme, by which *Desulfovibrio* reduces sulfate to hydrogen sulfide. The organism thus participates directly in the corrosion process by consuming the monatomic layer of adsorbed elemental hydrogen atoms generated at cathodes.[8] Eqs. (2.6) and (2.7) represent the formation of corrosion products, and Eq. (2.8) is the net reaction, showing that the ratio of total iron corroded to the iron transformed to ferrous sulfide should be 4 : 1; in practice the ratio is usually less.

Still another factor in anaerobic corrosion is the direct attack of metallic iron by hydrogen sulfide.

$$H_2S + Fe = FeS + H_2 \tag{2.9}$$

The ferrous sulfide so formed is cathodic to metallic iron. Direct oxidation of hydrogen sulfide by chlorine yields hydrochloric acid, which lowers the pH locally; this probably is significant in the attack on concrete basins.

$$H_2S + Cl_2 = 2H^+ + 2Cl^- + S \tag{2.10}$$

Cooling water always contains an ample supply of sulfate ion to sustain the activities of sulfate-reducing bacteria. It is introduced in the make-up water, in sulfuric acid added to adjust pH, and in commercial, dry chemical formulations that contain sodium sulfate as an antidusting or anticaking agent.

Corrosion induced by anaerobic bacteria is uncommon in clean chromate-treated cooling systems. Drummond and Postgate[9] have reported that 2 ppm of chromate ion inhibits the growth of sulfate-reducing bacteria, but does not affect aerobes. Presumably chromate ion is inimical to the strongly reducing conditions required for the enzymes of anaerobes.

Clostridium nigrificans is another anaerobic sulfate-reducer native to sewage, soil, and mud, but it has not been definitely implicated in anaerobic corrosion. It is a spore-forming thermophile, growing well up to 55°C, and in the wide pH range of 5-9.

Crenothrix, *Leptothrix*, and *Gallionella*, the so-called iron-bacteria, are frequently mentioned, but seldom seen in cooling systems. They oxidize ferrous to ferric iron, which then precipitates as the hydrous oxide mixed with a mucilaginous secretion, producing a voluminous brown slime.[10] These bacteria may initiate pitting and tuberculation in iron pipes. They are partial to wells, or to outfalls where water is entering a stream, but are rare in recirculating cooling water for, as mentioned before, they do not find an environment with high oxygen tension congenial. Similarly, *Thiobacillus*, which metabolizes sulfur or hydrogen sulfide to sulfuric acid, cannot live in cooling water because it favors an environment in the pH range of 2.0-3.5, although it remains viable to pH 6.0.

3. ALGAE

Algae are often visible as green, felt-like mats in internal sections of cooling towers that are wet and accessible to sunlight. Although more obvious than bacteria, algae are generally less troublesome because their growth is relatively easy to control with microbicides. Tehle,[11] in examining the interior structures of a large number of cooling towers found four genera to be most prevalent: *Oscillatoria* (30 percent); *Chlorella* (28 percent); *Ulothrix* (24 percent); and *Scenedesmus* (14 percent). He reported a total of thirteen genera of green algae, eight of diatoms, and three of blue-green algae. In chromate-treated systems green algae and diatoms predominate.

Filamentous and capsulated colonial algae coat splash packing and interior structural members of cooling towers, thus interfering with the formation of droplets that is required for intimate contact between water and air. Algal growths also prevent the rain of water from wetting film packing and forming thin films. Both of these effects decrease the rate of heat rejection in an evaporative cooling tower. As the optimum temperature for respiration of *Chlorella* and *Scenedesmus* is 27°C, and that of *Oscillatoria* is 40°C, it would be expected that the latter would prefer the upper part of the structure where the water is warmer.

The United States Department of Health, Education, and Welfare[12] has published a manual on the identification and significance of algae that contains six striking color plates of a variety of these aquatic organisms. Algal plants cannot be differentiated by the naked eye, of course, but their identification is unequivocal when viewed under a microscope.

Although ordinarily less troublesome than capsulated bacteria, algae also secrete a mucilaginous sheath analogous to the slime layer in bacteria. In blue-green algae, especially, this mucilage is associated with a peculiar form of movement called gliding, that is exhibited when the organism is in contact with a firm surface. In the process of gliding a filament of *Oscillatoria* leaves behind a track of mucilage similar to that of a moving snail. The mucilage seems to be secreted through pores in one of the cells. Furthermore, these polysaccharide-based mucilages can be extracted from many species of algae with hot water. Another slime-producer is the diatom *Navicula*, which when lacking sufficient silicic acid for its cell wall, secretes quantities of mucilaginous capsular material.[13] Under some conditions these sticky secretions may contribute significantly to complications in cooling systems such as clogged screens and fouled heat exchangers.

Myers[14] has discussed the role of algae in corrosion, but this particular aspect is seldom important in recirculating cooling systems, although *Anacystis* has been known to attack concrete basins. Many algae produce within their cells malodorous oils and poisonous cyclic alcohols, notably furfuryl alcohol:

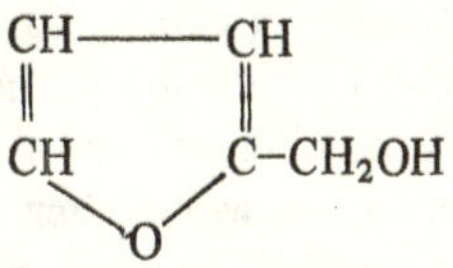

These are released when the organisms die, and may cause odor complaints unless they are killed with a suitable biocide. The capsulated colonies of the blue-green alga *Anacystis* are prone to generate septic odors, but this genus is infrequent in cooling towers; most of the common inhabitants are not objectionable. In the practice of water treatment there is constant reference to "dead algae," although as a practical matter there is no such thing. In common with other microorganisms, algae when they expire immediately undergo autolysis and shortly vanish. Lewin[4] reports that the relationship between dry weight and cell volume for many unicellular algae falls in the range of 0.22–0.27 g/ml, i.e., they are 73–78 percent water. When treated with a biocide they dry up and disappear. Slimy masses removed from heat transfer equipment more often than not are highly hydrated mixtures of dirt and hydrous oxides; here again, the microscope provides positive identification.

4. *FUNGI*

Cellulolytic fungi attack fibers of cellulose in wooden cooling towers, and if allowed to grow unchecked, can virtually destroy the structure. Wood is approximately 50 percent cellulose and hemicellulose, with about 30 percent of lignin, which serves as a binder and matrix for fibrils of cellulose. Redwood, fir, or pine are most often used for construction, primarily because they have favorable strength/weight ratios, as well as a certain amount of natural resistance to fungus infection. This resistance arises from the presence in these woods of 10–40 percent of certain cyclitols, tannins, and phenolics known collectively as "extractives," which being water-soluble are slowly leached from the wood over a period of time. When new redwood structures are put into service a condition called "black water" often persists for a few days as large amounts of dark-colored extractives are dissolved from the new lumber. Apparently these compounds are toxic to fungi, and thus provide natural protection for a few years. Their removal is accelerated, however, by pH values greater than 7.8 in the recirculating water, and by chlorine, both of which make the wood more vulnerable to invasion by fungi.

Wood in the flooded sections of a cooling tower is subjected to a continuous process of delignification, whereby the matrix of lignin is slowly dissolved leaving loose fibrils of cellulose exposed on the surface, and imparting a bleached appearance to the wood when dry. These surface fibers are gradually washed away, eventually lodging in the screens ahead of the circulating pumps, or else within the cooling system itself. This chemical attack is usually most obvious on the fill slats, but it also occurs to a lesser extent on air intake louvers and in certain parts of the plenum, such as the drift eliminators. High concentrations of dissolved solids can damage wood that is alternately wet and dry, as salts crystallizing in the dry areas rupture wood cells. Chemical erosion is entirely a surface effect so the strength of the wood is not appreciably affected for many years, but it is invariably a precursor of attack by soft-rot fungi.

Soft-rot typically attacks the surface of wood in flooded sections, and less frequently, in the plenum; often an extensive mycelial growth is visible having a characteristic mold-like appearance. *Fungi Imperfecti* are usually the cause of soft-rot, but a few genera of the *Ascomycetes* class have also been isolated on occasion from redwood cooling towers. In wood that is continuously wet fungal hyphae ramify through cell cavities, penetrating fibers and cells transversely, but if wood is wet only occasionally the hyphae grow longitudinally, following the fibrils of cellulose within the thick secondary wood cells. When dried the wood appears normal, but its surface is darkened, cracks develop across the grain, and it breaks with a brash fracture. Cellulose is depleted, but lignin is not attacked much.

Four species of the class *Basidiomycetes* are the most prevalent cellulolytic fungi in cooling towers, all of which produce an insidious internal decay, or pocket rot. *Poria negrescans* and *Peniophora mollis* cause white pocket-rot in heavier and drier structural members, cell partitions, decks, fan stacks, and drift eliminators. White-rot consumes much of the lignin as well as the cellulose, and here again, chlorine and alkaline solutions intensify the decay.

Brown-rot is caused by *Poria monticola* and *Poria oleaceae.* As these species have little tolerance for moisture or low oxygen tension, brown-rot is confined to relatively dry members of the structure. It is characterized by a rapid increase in the alkali-solubility of the infected wood, which is attributable to the local accumulation of acidic degradation products of cellulose. Lignin is not severely attacked, although the amount steadily decreases. Wood infected with either white- or brown-rot appears sound externally, but within there may be such extensive decay that the structure is dangerously weakened.

Other things being equal, redwood is attacked more slowly by soft-rot than by white- or brown-rot. Also, soft-rot fungi exhibit a higher tolerance to wood preservatives than do the *Basidiomycetes.* Many other fungi and bacteria are found in decayed wood, but their role has not been definitely established. A few molds and yeasts also appear now and then, but none of them is of any permanence or significance. Two outstanding contributions to the subject of wood deterioration by fungi have been published by

Duncan.(15,16)

This preliminary discussion is closed with a tabular summary of the predominant organisms, their usual order of prevalence, some of their significant characteristics, and their undesirable effects on the efficiency of cooling systems.

TABLE 2.1

Microorganisms in Aqueous Cooling Systems

Bacteria

Genus	Type	Usual order of prevalence	Spore-forming	Gram reaction	Undesirable effects
*Pseudomonas**	Aerobic	1	No	–	Slime
*Aerobacter**	Aerobic	2	No	–	Slime
Flavobacterium	Aerobic	3	No	–	Slime
Desulfovibrio	Sulfate-reducer	4	No	–	Pitting; odor
Bacillus	Anaerobic	5	Yes	+	Negligible
Clostridium	Sulfate-reducer	6	Yes	+	Cellulolytic, odor
Sarcina	Anaerobic	7	No	+	Negligible
Escherichia	Enteric	8	No	–	Negligible

Algae

Division	Genus	Type	Usual order of prevalence	pH range	Undesirable effects
Blue-green	*Oscillatoria**	Filamentous	1	6.0–9.5	Slime; coats fill
Green	*Chlorella**	Unicellular, colonial	2	4.5–9.3	Fouling
Green	*Ulothrix**	Filamentous	3	6.0–8.9	Coats fill
Diatom	*Melosira**	Filamentous	4	6.0–8.9	Fouling
Diatom	*Fragilaria*	Filamentous	5	6.0–8.9	Fouling
Green	*Scenedesmus*	Unicellular	6	6.0–7.9	Negligible
Diatom	*Navicula*	Unicellular	7	6.0–7.9	Slime
Blue-green	*Anacystis*	Capsulated-colonial	8	6.0–7.9	Slime; odor
Green	*Chlorococcum*	Unicellular-colonial	9	6.0–8.9	Slime

TABLE 2.1 *(cont'd.)*

Fungi				
Class	Genus or Species	Usual order of prevalence	Spore-forming	Undesirable effects
Basidiomycetes	*Poria negrescans**	1	Yes	White-rot
Basidiomycetes	*Peniophora mollis**	2	Yes	White-rot
Basidiomycetes	*Poria montricola**	3	Yes	Brown-rot
Basidiomycetes	*Poria oleraceae**	4	Yes	Brown-rot
Fungi Imperfecti	*Bisporomyces**	5	Yes	Soft-rot
Fungi Imperfecti	*Phialophora*	6	Yes	Soft-rot
Fungi Imperfecti	*Cytosporella*	7	Yes	Soft-rot
Ascomycetes	*Chaetomium*	8	Yes	Soft-rot

* Predominant genera or species in recirculating cooling systems.

References

(1) Lewis, R. F. 1965. Control of sulfate-reducing bacteria. *J. Am. Water Works Assoc.* 57:1011.
(2) Lochhead, A. G. 1952. The nutritional classification of soil bacteria. *Proc. Soc. Appl. Bact.* 15:15.
(3) Lochhead, A. G. and Thexton, R. H. 1952. Qualitative studies of soil microorganisms. X. bacteria requiring vitamin B_{12} as growth factor. *J. Bact.* 63:219.
(4) Lewin, R. A. 1962. *Physiology and biochemistry of algae.* New York and London: Academic Press.
(5) Costello, J. A. 1969. The corrosion of metals by microorganisms. A literature survey. *Int. Biodetn. Bull.* 5:3, 101.
(6) Iverson, W. P. Mar., Apr. 1967. A possible role for sulfate-reducers in the corrosion of aluminum alloys. *Electrochemical Technology* 5:3–4, 77.
(7) Starkey, R. L. 1956. Transformation of sulfur by microorganisms. *Ind. Eng. Chem.* 48:1429.
(8) Hedrick, H. G., Crum, M. G., Reynolds, R. J. and Culver, S. C. Mar., Apr. 1967. Mechanism of microbiological corrosion of aluminum alloys. *Electrochemical Technology* 5:3–4, 75.
(9) Drummond, J. P. M. and Postgate, J. R. 1955. A note on the enumeration of sulfate-reducing bacteria in polluted water and on their inhibition by chromate. *J. Appl. Bact.* 18:307.
(10) Starkey, R. L. 1945. Transformations of iron by bacteria in water. *J. Am. Water Works Assoc.* 37:963.
(11) Tehle, E. 1961. Here's what grows in cooling tower water. *Oil Gas J.* 59:3, 84.
(12) Palmer, C. M. 1962. Algae in water supplies. U. S. Department of Health, Education, and Welfare, Public Health Service Publication No. 657. Washington, D. C.
(13) Lewin, J. C. 1955. The capsule of the diatom *Navicula pelliculosa. J. Gen. Microbiol.* 13:162.
(14) Myers, H. C. 1947. The role of algae in corrosion. *J. Am. Water Works Assoc.* 39:322.
(15) Duncan, C. G. 1960. Wood-attacking capacities and physiology of soft-rot fungi. Forest Products Laboratory, Forest Service, U. S. Dept. Agriculture, Report No. 2173.
(16) Duncan, C. G. 1960. Soft-rot in wood, and toxicity studies on causal fungi. *Proc. Amer. Wood-Preservers' Assoc.* 56:27.

III.

The Microbial Community

The microorganisms commonly encountered in open recirculating cooling systems have been described. Now a number of variables that affect the growth and health of this microbial community in this unique environment are considered.

1. FACTORS AFFECTING POPULATIONS

Both algae and bacteria are insensitive to pH changes through the range normally prescribed (6.0-8.0) for various cooling water treatments. Optimum respiration is achieved by many organisms, however, at specific values of pH and temperature. Thus, for blue-green algae the optimum pH for respiration is about 8. On the other hand, the green alga, *Chlorella*, is indifferent to pH between 4.5 and 9.3. Some algae assimilate phosphate most rapidly between pH 6 and 7, which includes the range commonly specified for phosphate-based treatments. The rot-producing fungi grow optimally near pH 6, and are affected by relatively small changes in pH. Thus, *Basidiomycetes* is inhibited by the mildly alkaline environment at pH 7-8, although in common with most other fungi, they remain viable between pH 2 and 8.

Respiration is also affected by temperature. The bacterial genera most numerous in cooling water, *Pseudomonas* and *Aerobacter*, prefer temperatures of 37 C and 34 C, respectively. Among the common algae, *Oscillatoria* grows best at 40 C, and *Chlorella* at 27 C, whereas the diatom *Melosira* thrives in the broad range of 18 to 35 C. A few green algae in soil even flourish above 60 C, for example in hot springs. The rate of respiration of algae is also influenced by the intensity of incident light, and by the concentration of carbon dioxide in the atmosphere. Many genera of fungi prefer temperatures in the range of 25–30 C, but for those that infest cooling tower wood the optimum temperature is high–about 34 C on the average.

Chromate ion, an efficient anodic corrosion inhibitor, is used in many cooling systems. In addition to its principal function, it has the added advantage of exerting a leveling effect on the growth of the microbial community; it is especially toxic to anaerobic organisms.[1] Zinc ion, which is customarily used along with chromate to provide cathodic protection, is itself inhibiting to some microorganisms. The concentration of zinc ion that is normally in solution—about 0.5 to 2 ppm—is not actually toxic, but it contributes to the leveling effect of chromate. Conversely, some chemicals used to inhibit corrosion, scaling, and fouling in cooling systems promote the growth of bacteria, thus making microbiological control more difficult. Nonchromate treatments, in general, favor the growth and multiplication of microorganisms because often the inhibitors used are themselves biodegradable, which means, of course, that microorganisms can use them to support respiration or growth. Also, some biocides are ineffective under the alkaline conditions that usually prevail in these treatments, a matter that is considered in Chapter V.

Several compounds containing phosphorus are used in water treatment for preventing calcium scales. Most of these, including polyphosphates, phosphonates, and phosphate esters, function as antiprecipitants of calcium carbonate and calcium sulfate.[2] Phosphorus is pentavalent in all of these compounds; in the polyphosphates the phosphate groups are tetrahedrally coordinated to shared oxygen atoms.

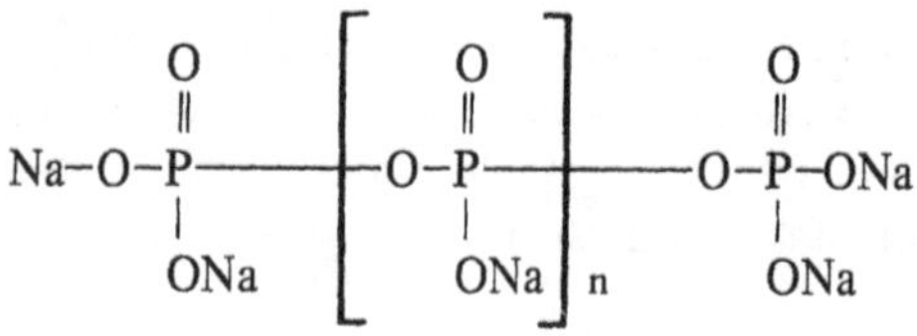

When dissolved in water these salts slowly hydrolyze to ortho-phosphate, the rate increasing with increasing temperature, decreasing pH, and increasing calcium concentration.

$$Na_2PO_3(NaPO_3)_nOPO_3Na_2 + (n + 1)H_2O = (n + 2)HPO_4^{--} + (n + 4)Na^+ + nH^+ \quad (3.1)$$

Phosphate esters typically contain the group

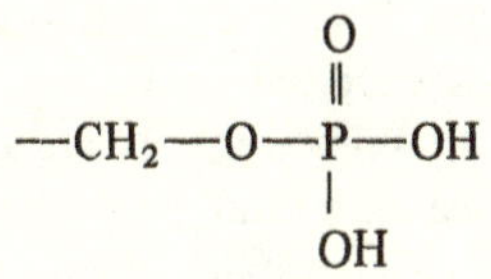

in which the carbon-oxygen bond cleaves readily, also forming ortho-phosphate as well as an organic fragment. The phosphonates, however, contain exceptionally stable carbon-phosphorus bonds, and so do not hydrolyze to ortho-phosphate in normal use, nor indeed, in boiling dilute sulfuric acid. The two phosphonates most often used to treat cooling water are amino-methylenephosphonate,

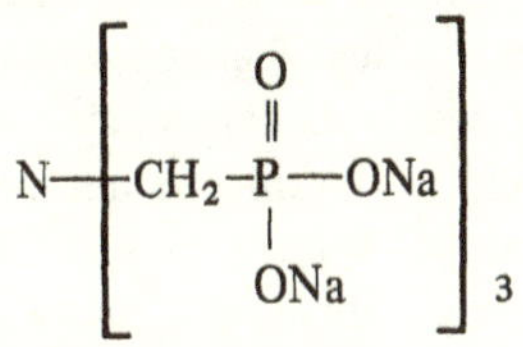

and hydroxyethylidenediphosphonate,

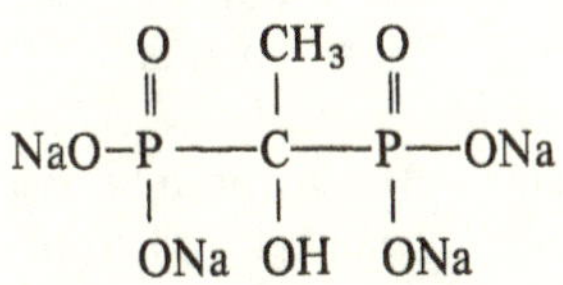

Compounds that yield ortho-phosphate on hydrolysis thereby provide an essential nutrient for microorganisms, which can assimilate phosphate only in the ortho form. Ortho-phosphate is dehydrated to polyphosphate by many organisms, and stored in volutin granules within their cells. This occurs in bacteria and fungi as well as in algae. For this reason the control of microorganisms, especially algae, is likely to be more difficult in systems treated with polyphosphates or phosphate esters than in those treated with phosphonates, or some other antiprecipitant, such as polyacrylate.

Lignins are still used to a limited extent to treat cooling water, often combined with polyphosphates or phosphonates. As *Flavobacterium* and some pseudomonads metabolize lignins to humic acids, high bacteria counts can be expected in systems using this chemical, especially in combination with polyphosphate. Starch derivatives used as binders in some solid formulations also can stimulate bacterial growth.

Commercial cleaners containing polyacrylate and chelant are sometimes advocated for dissolving calcium and iron phosphates in heat transfer equipment. The chelant is either ethylenediaminetetraacetic acid (EDTA), or nitrilotriacetic acid (NTA), also called triglycollamic acid. The effect of these chelants on the microbial community is ambiguous on account of antithetical reactions that can affect metabolism. These chemicals form stable chelates with cations such as copper and zinc thereby enabling microorganisms, algae in particular, to grow in the presence of otherwise toxic concentrations of these metal ions. Furthermore, these chelates solubilize and prevent the precipitation of phosphate that would otherwise be unavailable as a nutrient. On the other hand, essential trace metals such as magnesium and manganese, when chelated, are unavailable to microbes, which accordingly suffer a critical deficiency. Some algae, however, are able to biologically degrade EDTA and obtain metals from their chelates; *Scenedesmus* is one of these.[3]

A substantial amount of the dirt drawn into a cooling tower collects on the bottom of the basin, and with the passage of time, as oxygen is depleted, favorable conditions develop for anaerobic bacteria and a few other organisms. *Desulfovibrio* becomes active when the accumulation of mud becomes anaerobic. This may lead to odors and a certain amount of black ferrous sulfide in the basin. Davis[4] points out that sulfate-reducing bacteria often become active after aerobic bacteria consume all of the available oxygen, or produce organic wastes that can be used by anaerobes. A community of *Protozoa*, *Nemathelminthes*, and *Rotifera* is not uncommon on the surface of the mud at the bottom of a cooling tower basin. All of these animals occasionally appear in other places where stagnant water may be found, as for example, in water-jackets of compressors, or in water-cooled equipment that has been out of service for a while. *Protozoa* and *Rotifera* consume organic debris, and feed voraciously on bacteria and unicellular algae, thus keeping the population of microorganisms small in these particular locations. Harrison and Heukelekian[5] regard the bacteria-consuming rhizopods as accessory organisms, or secondary invaders of what was originally a bacterial infestation.

The effect of process leakage on the population of microorganisms in cooling water has been discussed in a general way in a great number of publications. It seems to be generally supposed that when a cooling system becomes contaminated with chemicals such as ammonia, ethanolamines, alcohols, urea, petroleum, etc., an enormous population of bacteria is immediately engendered that generates copious quantities of microbiological debris. It is well-known, however, that microorganisms proliferate in the presence of a single contaminant only if it is usable as a nutrient by the microorganisms, and if all other growth factors are present. Small leaks of chemicals yielding concentrations of a few parts per million in the recirculating water are seldom significant, as the low surface to volume ratio of bacteria does not favor the absorption of organic substances from dilute solution. Lincoln and Foster[(6)] in experiments dealing with river pollution found that a minimum concentration of 50 ppm of sulfite digestor liquor was necessary to produce any visible growth of *Sphaerotilis natans* after seven days; above this concentration the amount of growth was proportional to concentration. Furthermore, the intracellular tension in bacteria ranges from 6-30 atm., so that concentrations up to several hundred parts per million may be required to surmount this osmotic barrier, and move through the cytoplasmic membrane. Monoethanolamine at less than 20 ppm has little effect, but at concentrations of 100-500 ppm the number of bacteria eventually increases to very high values, i.e., 1.0×10^7-2.0×10^7/ml. Ammonia up to 200 ppm has little effect, as do most petroleum products up to about 20 ppm. The effect of contaminants on microorganisms is discussed further in Chapter VII.

Genera that inhabit cooling water are of necessity nutritionally versatile, particularly *Pseudomonas*, but no organism that has been subsisting on organic matter in soil can suddenly begin to metabolize a foreign material such as paraffins, amines, or cyclic hydrocarbons. A few members of any large bacterial community, however, may be able to metabolize one of these foreign substances, in which event their number increases, eventually shifting the balance of the population. That is, if special nutrients are constantly present, genera that can use them will proliferate. Spore-forming bacteria, for instance, which normally are scarce, predominate in cooling systems subjected to continuous leaks of ammonia, monoethanolamine, or glycols; fungi also are favored by such leaks. If nitrite is present, as it often is in cooling systems of plants manufacturing nitric acid or fertilizer, *Nitrobacter agilis* develops. Urea can serve as the sole nitrogen source for *Chlorella* and some other unicellular green algae, but its ability to increase the permeability of the cytoplasmic membrane makes it toxic to others, such as *Spirogyra*.

The osmotic pressure in algal cells is considerably less than that in most bacteria so lesser concentrations of certain contaminants may enhance the growth of algae. Thus, *Chlorella* can oxidize small amounts of salts of organic acids, including acetate, fumarate, and succinate to carbon dioxide. Intracellular pressure ranges from 4-8 atm. in diatoms, and is 7-8 atom. in the common green and blue-green algae.

Petroleum leaking into cooling water seldom increases the number of bacteria because species that can subsist on petroleum are normally not present. After a prolonged period of contamination, however, petroleum-degrading organisms do appear. In general, the more soluble the petroleum fraction is in water, the more easily it is metabolized. McAuliffe[7] has measured the solubilities in water at room temperature of 65 hydrocarbons, and as might be expected, his results show that solubility varies with structure and molecular weight of the hydrocarbons. The solubility of ortho-xylene, for example, is about 175 ppm, while that of normal octane is about 0.7 ppm.

When water and petroleum are present in two phases virtually all of the viable microorganisms are in the water, as they cannot survive in petroleum. Hill, et al.[8] suggest that the reason for this may be that cellular lipids are soluble in hydrocarbons. Indeed, Albro and Huston[9] have found that 20 percent of the normal cellular lipids synthesized by *Sarcina lutea* are normal paraffins in the C_{23}-C_{31} range. As a rule, paraffins are most easily metabolized by *Pseudomonas*; chromatograms of petroleum fractions extracted from cooling water contaminated for some time by oil show a considerable diminution of elution peaks produced by paraffinic chain structures. Less readily, *Pseudomonas* can also use naphthalene, phenanthrene, benzene, toluene, and xylenes; it is also one of the primary organisms implicated in the degradation of jet fuel.[8]

Pseudomonas methanitrificans is a gum-forming species that fixes nitrogen from air, and oxidizes methane to a waxy substance composed of the microorganisms themselves and their metabolic products, such as humic acids.[4] Sulfate-reducing bacteria anaerobically metabolize petroleum, including some crude oils, but at a much slower rate than do the aerobes. *Desulfovibrio* also requires adequate supplies of nitrogen, phosphate, and sulfate, emphasizing again that a single contaminant, even if usable as a source of carbon, must be accompanied by all essential growth factors before any significant increase in the number of organisms can occur.

2. *GROWTH RATES*

Bacterial growth results from the metabolic decomposition of organic compounds into fragments suitable for the synthesis of new cells. Bacteria reproduce by binary fission, a process whereby the number of cells doubles with each generation; the time interval required for each doubling is the generation time. Thus, if n is the number of intervals and N is the total number of bacteria, then

$$N = 2^n \tag{3.2}$$

Under optimum conditions a bacterium divides into two every 20 minutes. Were this rate maintained a single organism ($n = 0$) would give over one billion new cells ($n = 30$) in a period of ten hours. This exponential growth cannot continue long, however, because nutrients are exhausted, toxic metabolic wastes accumulate, physiological crowding ensues, and many cells die. If the number of organisms at time, t_1, is N_1 and the number at t_2 is N_2, after g generations, then $N_2 = 2^g N_1$. Taking logarithms to base 2, g is evaluated as

$$g = \log_2 N_2 - \log_2 N_1 \tag{3.3}$$

The number of generations per time interval, $(t_2 - t_1)$, is $\mu(t_2 - t_1)$, where μ is the number per unit time, i.e., the specific growth rate. Thus,

$$N_2 = 2^{\mu(t_2 - t_1)} N_1 \tag{3.4}$$

Or

$$\log_e N_2 = \mu(t_2 - t_1)\log_e 2 + \log_e N_1 \tag{3.5}$$

from which

$$\mu = \frac{\log_e N_2 - \log_e N_1}{(t_2 - t_1)\log_e 2} \tag{3.6}$$

If $\log_e N_2$ is plotted against t an exponential curve is obtained of the type shown as B in Fig. 3.1. Initially, in the lag phase, the rate of growth is slow, after which, in the logarithmic phase, $d(\log_e N)/dt$ attains its maximum value. A stationary state is then reached with the maximum number of bacteria that the medium can support, and finally comes the death phase, in which the organisms begin to die, again at a logarithmic rate. This growth pattern is typical of a closed culture; that is, a definite volume of culture medium containing an inoculum of bacteria. In a recirculating cooling system the water is the culture medium, but it is not closed, for water is continuously withdrawn in the blowdown, and additional bacteria and nutrients are constantly being transferred to the water from dust in the air drawn through the cooling tower.

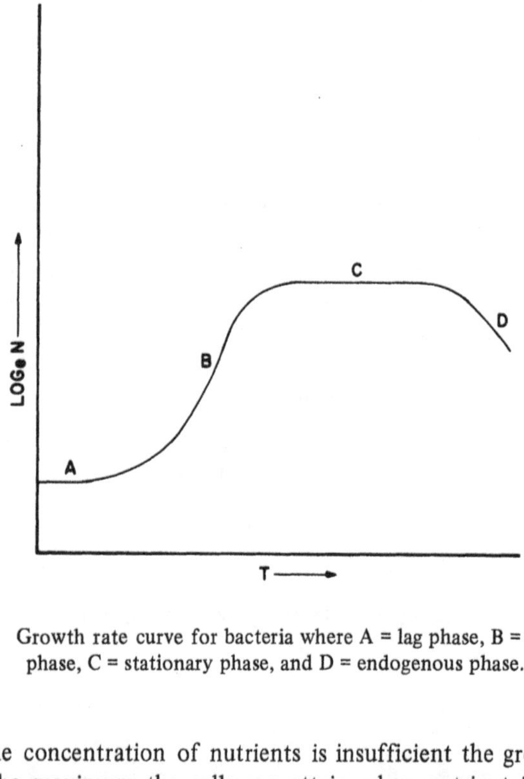

Fig. 3.1 Growth rate curve for bacteria where A = lag phase, B = logarithmic phase, C = stationary phase, and D = endogenous phase.

When the concentration of nutrients is insufficient the growth rate is less than the maximum the cells can attain when nutrient is increased. Cooling water at best is nutritionally deficient, and the bacteria it contains are in a continuous state of virtual starvation, so that in the absence of some sort of usable contamination, the maximum number of bacteria that can live per milliliter of cooling water is about 2.0×10^6. In the presence of easily metabolized chemicals such as monoethanolamine or glycols, however, the maximum appears to be about 10^7. On the other hand, the growth rate may be limited by a lack of essential growth factors, or by some unknown condition within the cells rather than by the concentration of a supplied nutrient, and indeed, in many instances process leaks have no discernible effect on the growth rate of the microorganisms in a cooling system.

Hughes[10] has given the approximate weight of a bacterial cell as 10^{-12} g, that is, 10^9 bacteria weigh about one milligram. It is instructive to calculate the total weight of bacteria in a cooling system of, say, 200,000 gal when the bacterial count is 10^6/ml.

$$\begin{aligned}
\text{bacteria/ml} &= 1 \times 10^6 \\
\text{bacteria/mg} &= 1 \times 10^9 \\
\text{volume of water} &= 200{,}000 \text{ gal} \\
&= 3785 \times 200{,}000 \text{ ml} \\
&= 7.57 \times 10^8 \text{ ml} \\
\text{total bacteria in system} &= 7.57 \times 10^8 \times 1 \times 10^6 \\
&= 7.57 \times 10^{14} \\
\text{mg bacteria in system} &= 7.57 \times 10^{14}/10^9 \\
&= 7.57 \times 10^5 \text{ mg} \\
\text{lb bacteria in system} &= 7.57 \times 10^5/454 \times 10^3 \\
&= 1.67 \text{ lb } (1.0 \text{ ppm})
\end{aligned}$$

The point of this is that whether bacteria are swimming in the water, or lying somewhere in the cooling system they must have something to metabolize in order to live and multiply. As mentioned above, the number that can be nourished by clean cooling water is limited, and it is for this reason that relatively few bacteria are seen in the majority of fouling deposits when they are examined under the microscope.

Growth rate is also markedly affected by temperature: at 10–12 C it is slow; for many bacteria 30–40 C is optimum; above 45 C the growth rate slows again. Between 50 and 70 C thermal death of most microorganisms occurs because of the rapid destruction of heat-labile proteins and nucleic acids. At 25 C growth is faster than at 15 C by a factor of three, thus microbiological control tends to be easier in winter than in summer, although complicated to some extent by the tendency of some bacteria to produce voluminous capsules at low temperature. It should also be pointed out that pseudomonads are psychrophilic, and grow remarkably well at temperatures approaching 0 C. In Table 3.1 are the generation times, t_2, of *Escherichia coli* at different temperatures, as given by Salle.[11]

TABLE 3.1

Effect of Temperature on Generation Time

°F	°C	t_2
50	10	860 min
59	15	120 min
68	20	90 min
77	25	40 min
86	30	29 min
95	35	22 min
104	40	17.5 min
113	45	20 min
118	47.5	77 min

Cell division in algae is much slower than in bacteria (one to three divisions per day on the average) and some of them have specific requirements, including amino acids, certain essential cations, and various vitamins. As the vitamins are mostly of bacterial origin sudden algal blooms may follow an increase in bacterial growth. Conversely, cultures of *Chlorella pyrenoidosa* have been identified by Spoehr, et al.,[(12)] as oxidation products of unsaturated fatty acids, which have pronounced bactericidal properties. *Oscillatoria splendida* also secretes bacteriostatic substances. This may account for the observation that dense algal blooms in cooling towers are sometimes accompanied by a decrease in the number of bacteria in the recirculating water. Then again, the bacteria may merely adhere to the sticky surface of the algal mass, depleting the number in the water.

References

(1) Drummond, J. P. M. and Postgate, J. R. 1955. A note on the enumeration of sulfate-reducing bacteria in polluted water and on their inhibition by chromate. *J. Appl. Bacteriol.* 18:307.

(2) McCoy, J. W. 1974. *The chemical treatment of cooling water.* New York: Chemical Publishing Company.

(3) Lewin, R. A. ed. 1962. *Physiology and biochemistry of algae.* New York and London: Academic Press.

(4) Davis, J. B. 1967. *Petroleum microbiology.* New York: Elsevier.

(5) Harrison, M. E. and Heukelekian, H. 1958. Slime infestation—literature review. I. *Sphaerotilus. Sewage Ind. Wastes* 30:1278.

(6) Lincoln, J. H. and Foster, R. F. 1943. Investigation of pollution in the lower columbia river. Washington State Pollution Commission and Oregon State Sanitary Authority.

(7) McAuliffe, C. 1966. Solubility in water of paraffin, cycloparaffin, olefin, acetylene, cycloolefin, and aromatic hydrocarbons. *J. Phys. Chem.* 70:1267.

(8) Hill, E. C., Evans, D. A., and Davies, I. 1967. The growth and survival of microorganisms in aviation kerosine. *J. Inst. Petrol.* 53:280.

(9) Albro, P. W. and Huston, C. K. 1964. Lipids of *Sarcina lutea.* II. hydrocarbon content of lipid extracts. *J. Bacteriol.* 88:981.

(10) Hughes, D. E. *Microbiology.* 1968. Institute of Petroleum, London.

(11) Salle, A. J. 1973. *Fundamental principles of bacteriology.* 7th ed., New York: McGraw-Hill.

(12) Spoehr, H. A., Smith, J. H. C., Strain, H. H., Milner, H. W., and Hardin, G. J. 1949. Fatty acid antibacterials from plants. Carnegie Inst. Wash. Pub. No. 586, 67.

IV.

The Microbiology of Reclaimed Water

The increasing urgency to conserve water and reduce pollution has led to the use of treated waste water for make-up to cooling systems, often with their blowdown being returned to the treating plant for recycling. This type of water introduces microorganisms into cooling systems along with residual organic matter, which fosters microbial growth, and therefore, for reasons discussed in Chapter VI, Section 4, it is usually necessary to curtail cycles of concentration. Because of the increasing reuse of industrial process water for cooling, it is of interest to examine some aspects of the practice paying particular attention to possible adverse effects within operating cooling systems. Before this can be done, however, it is necessary to review a few of the methods of treating industrial wastes; these include biological oxidation, trickling filtration (fixed film biological reaction), aerated lagoons, oxidation ponds, photosynthetic processes, composting, and anaerobic digestion. We will consider three of these processes in sufficient detail to anticipate difficulties to be expected when their effluents are used as make-up for recirculating cooling systems. The processes are: activated sludge, the rotating biological contactor (fixed film), and waste stabilization ponds. All of these are methods of secondary sewage treatment that can also be applied more-or-less successfully to the treatment of industrial wastes.

There is a voluminous, though largely qualitative, literature devoted to the treatment of domestic sewage, but because of their great variety, information on methods of processing industrial wastes is relatively meager. Nevertheless, it is readily apparent that effluents from such diverse industries as poultry processing, pulp and paper manufacturing, distilleries, chemical processing, and petroleum refining, offer a considerable challenge to the metabolic versatility of microorganisms. Although the chemical composition of specific industrial wastes can usually be more precisely characterized than that of domestic sewage, the latter is, with few exceptions a fairly predictable mixture. Table 4.1 lists approximate composition ranges of the two types of waste.

TABLE 4.1

Characteristics of Wastes

Component	*Domestic Sewage*	*Industrial Waste*
BOD*, ppm	150–300	⩽ 50,000
COD**, ppm	250–500	⩽ 100,000
TOC†, ppm	100–150	⩽ 25,000
Nitrogen, ppm N	20–40	3–5000
Phosphorus, ppm P	3–12	1–100
Inorganic dissolved solids, ppm	300–800	⩽ 25,000
Suspended solids, ppm	100–300	0–5000
Alkalinity, ppm $CaCO_3$	50–200	highly variable
pH	6–7	2–12

*Biochemical oxygen demand
**Chemical oxygen demand
†Total organic carbon

Primary treatment includes coarse screening to remove a large proportion of solids, followed by skimming tanks (in petroleum refineries API separators or flotation units to remove undissolved oil) and sedimentation basins wherein further separation of solids takes place. Tertiary treatment is seldom used for industrial effluents, though some municipalities remove residual material that exerts a chemical oxygen demand by treating secondary effluent with activated carbon, then sell this so-called reclaimed water for industrial use. Secondary treatment depends upon the capacity of microorganisms to assimilate organic material after it has been biologically degraded. The three processes to be discussed in this chapter all depend on the metabolic activity of microorganisms, although mechanical features differ in each. Engineering aspects are largely ignored here, the emphasis being on microbiological considerations, especially as they affect subsequent reuse of the effluent water for cooling. The reader interested in operational details of sewage plants should consult the excellent treatment of this subject by Parker.(1)

1. *THE ACTIVATED SLUDGE PROCESS*

This process entails the transformation of organic waste to activated floc, at the same time providing a classic example of a continuous culture or steady-state microbial ecosystem. It is the most efficient (and expensive) method of purifying an amenable waste stream. The process exploits the high oxidation efficiency of microorganisms in their declining growth phase (see page 37), and their enhanced capacity to flocculate when in the endogenous stage of respiration, in which condition stored food is being consumed and oxygen consumption is reduced. Although studied more than any other process, little has been published on the treatment of industrial wastes by activated sludge. The process is considered by some to be better suited for the treatment of domestic sewage because microorganisms are especially sensitive to shock loading, that is, sudden changes in concentrations of waste material, altered flow rates, the occurrence of chemical spills, variations in pH, or slugs of toxicant, and also because many industrial wastes are not at all readily biodegradable. The concentration of a waste stream, for instance, can be changed abruptly by the temporary shutdown or startup of a processing plant, or by the influx of water from storm drains. Any of these events exerts stress on the microbial community.

As pointed out by Pipes,[(2)] the activated sludge process is carried out under a number of operational variations, but fundamentally influent water together with air is pumped into an aeration vessel at a rate (0.5–2.0 ft^3 of air per gallon of waste) sufficient to maintain a concentration of at least 1 ppm of dissolved oxygen in the mixed liquor. Under the influence of oxygen and the organic waste substrate, a diversified population of microorganisms develops and agglomerates into discrete particles of floc. The organic waste is oxidized to provide energy for the microorganisms, which synthesize cellular material from the products of metabolism. In a properly functioning treatment, the average length of time that a particle of suspended solids remains under aeration is about three and one-half days. All of the suspended solids, including microorganisms and organic and inorganic waste, are collectively called activated sludge. A normal particle is 50–500 μ in diameter and consists primarily of flocculated bacteria with short filaments projecting from each particle of floc. Upon aging there is an increase in the number of filamentous microorganisms in the floc particles, which for reasons to be discussed presently, is undesirable.

After aeration and a suitable period of retention in the aeration tank, the sludge is transferred to a sedimentation basin where the floc is allowed to settle for about 30 minutes. Satisfactory settling is a primary requisite for the successful operation of an activated sludge process, for if the sludge is not cleanly disengaged from the mixed liquor, organisms in the sludge increase the biochemical oxygen demand of the effluent by consuming oxygen. After settling, a portion of the sludge (perhaps 10 percent) is returned to the aeration tank and mixed with the incoming waste; the bulk of the settled solids is discarded.

To some extent the operation of an activated sludge plant is empirical, but there are a few controls. A waste stream is characterized by its biochemical oxygen demand (BOD), its chemical oxygen demand (COD), and the concentration of total organic carbon (TOC). The ratio (BOD/COD) is a measure of the fraction of total organic material that is capable of being biologically oxidized. Other special determinations such as organic acids, or volatile organics (alcohols, ketones, aldehydes, hydrocarbons) are of importance in the treatment of wastes from certain industries. For routine control daily measurements of pH, dissolved oxygen, suspended solids in mixed liquor, and sludge volume index usually suffice.

The efficiency of biological oxidation is estimated by comparing the five-day BOD of the influent with that of the effluent from the treatment plant. The reduction of BOD in domestic sewage ranges from 75-95 percent; in industrial waste water, however, the range is much wider. The degree of sedimentation is measured by timing the rate of settling of a portion of mixed liquor in a graduated cylinder. Other operational details have been given by Pipes.[2]

Three biological requirements must be satisfied to make the activated sludge process workable. First, a mixed community of aerobic microorganisms must be established that is collectively able to degrade a substantial proportion of the organic components in the waste stream to be treated. These organisms can be classified by function as floc-formers, saprophytes, predators, and nuisances. Floc-formers agglomerate particles consisting of microorganisms, organic debris, and inorganic material; they are vitally concerned in the formation of settleable floc. Saprophytes are organisms that feed on nonliving organic matter, thereby degrading whatever portion of waste they are able to metabolize. Predators include *Protozoa* and fungi that prey on the saprophytes; they may also have an effect on the formation of floc. Nuisance organisms are those that interfere with floc formation and settling of sludge. Filamentous bacteria, fungi, and zoogloeal bacteria are all undesirable for this reason. All of the organisms mentioned are continuously inoculated into the aeration vessel in water, soil, and sewage, so in the presence of a sufficiency of oxygen and organic substrate, biological oxidation is self-starting and self-sustaining. Some of the organisms feed on particulate matter, while others metabolize only soluble compounds. In one study,[3] less than one-third of the organic carbon and slightly more than one-third of the organic nitrogen were found to be in solution. Presumably the microbial population shifts to accomodate the food supply available.

The second requirement is that the useful microorganisms find the environment of the aeration tank congenial. Thus, conditions here must be such that the growth of desirable organisms is encouraged, while that of others is inhibited. It is difficult to be quantitative about this, but low pH, for example, favors the growth of fungi, particularly *Geotrichum*, that interfere with the formation of settleable floc. Furthermore, if nitrogen and phosphorus are deficient, carbonaceous material cannot be oxidized, and the growth of filamentous microorganisms is encouraged. In a nitrogen-deficient medium fungi predominate; if oxygen is inadequate *Protozoa* succumb. Finally, other nutritional elements such as calcium, iron, magnesium, manganese, potassium, sodium, chloride, and sulfate must be present in sufficient concentrations to support the oxidation process.

The third biological essential is that microorganisms be at the proper stage of growth to form floc that settles rapidly and completely in the sedimentation basin. This means that they must be in the endogenous stage, in which natural coagulants are secreted and capsules are elaborated in a form that favors coagulation. Under normal operating conditions only about 30 percent of the microorganisms in activated sludge are really active.

As in soil, Gram-negative genera of bacteria predominate in activated sludge, the more numerous being *Achromobacter*, *Flavobacterium*, and the ubiquitous *Pseudomonas*. Other genera are *Bacillus*, *Aerobacter*, *Sarcina*, *Alcaligenes*, *Spirillum*, and *Zoogloea*. Intestinal bacteria are present in domestic sewage, but are of little significance in the process. The filamentous bacteria *Sphaerotilus natans*, *Bacillus*, and possibly *Beggiatoa* are often encountered, but are undesirables as they too interfere with the settling of floc. Although *Beggiatoa* is rare in activated sludge, it is frequently encountered on trickling filters and in polluted streams. It grows in a narrow pH range around neutrality on simple, soluble substrates, but cannot metabolize complex materials; ammonia is a suitable source of nitrogen. The bacterium is capable of oxidizing moderate amounts of hydrogen sulfide, depositing elemental sulfur within the cell, but is poisoned by high concentrations of the gas. The organisms frequently exist as trichomes that occasionally break up into short filaments.

A variety of fungi, molds, and yeasts inhabit the mixed liquor in aeration tanks. A few of them are bacterial predators and all of them increase the filamentous character of floc, thus interfering with sedimentation. The following genera have been reported: *Mucor*, *Rhizopus*, *Aspergillus*, *Fusarium*, *Trichoderma*, and *Phialophora*. The latter, incidentally, causes soft-rot in wooden cooling towers.

Holozoic (solid food-eating) *Protozoa*, particularly ciliates, exert a beneficial effect by scavenging unflocculated unicellular bacteria, yeast cells, and particles of debris not entangled in the floc, in this way clarifying the effluent from the sedimentation basin. In general, *Protozoa* do not compete with bacteria for organic material, but some of them aid flocculation by secreting a mucus or a polysaccharide. Well over 200 species of *Protozoa* have been identified in activated sludge plants, 160 of which are free-swimming, crawling, or stalked ciliates. In addition, there are rhizopods and flagellates; among the invertebrates are rotifers and occasionally nematodes. As *Protozoa* are more vulnerable to toxicants than are bacteria, a toxic shock is often evidenced by increased turbidity in the effluent from the sedimentation basin. Also, these organisms do not survive temperatures exceeding approximately 36 C (97 F). The normal concentration of *Protozoa* in the mixed liquor is about 50,000 cells/ml. Algae do not grow in activated sludge.

Pipes[4] has criticized the emphasis in waste treatment on engineering design to the neglect of fundamental biological research, and he has summarized the requirements for efficient operation of the activated sludge process. Microbiological activity must be such that organic matter is removed from the waste stream in a reasonable length of time. The sludge must separate from the treater effluent by sedimentation. Toxicants must be absent, or present at innocuous concentrations. Finally, successful operation depends to a great extent on the proportions of nitrogen and phosphorus (phosphate) with respect to total organic material. As a general rule, phosphorus should amount to 0.6–1 percent, and nitrogen should be 3–6 percent of the BOD. Domestic sewage normally contains ample concentrations of both nitrogen and phosphorus, but industrial wastes are likely to be deficient in these elements, in which case they must be supplied.

The most crucial element in secondary treatment is to ensure that sludge is formed that settles rapidly and completely. Several variables influence the formation of suitable sludge. The presence of zoogloeal bacteria, which exist as individual cells embedded in a hydrous, gelatinous matrix, aid in the coagulation of floc by adhering to extracellular material produced by fungi and to particles of miscellaneous debris. Individual bacterial cells are self-flocculating in some circumstances, and it is now thought that *Protozoa* may play a role in the flocculation of some bacteria. As in natural surroundings, microorganisms exist in mixed liquor as individual cells, as aggregates of cells, and as filaments. If the concentration of soluble organic material becomes excessive in a waste stream, dispersed microbiological growth ensues, flocculation is inhibited, and little or no settleable sludge is formed. Because of this effect, the practical upper limit of concentration of organic matter (BOD) treatable by the activated sludge process is about 1000 ppm. As suggested before, a deficiency of nitrogen decreases the reduction of BOD, and slows the rate of growth of sludge. If the concentration of phosphate in the mixed liquor is less than about 5 ppm, the growth of filamentous organisms is favored over that of the floc-formers.

Activated sludge is insensitive to pH through the range of 6 to 9. Outside this range, however, especially below 5 and above 10, the microbiological population is altered so that normal sludge does not form. Sudden influxes of low-pH water cause sludge to disperse; if the condition continues filamentous organisms eventually predominate. In addition, the proliferation of ciliates and rotifers depletes saprophytic bacteria, leading to poor settling. Although remarkably resistant to some toxicants, activated sludge is adversely affected by toxic shocks. That is, sudden high concentrations of phenolics, cyanide, heavy metals, or microbicides are certain to damage sludge already formed, and can inhibit the coagulation of new floc particles. Similarly, abrupt increases in BOD caused by spills or intentional dumping of organics that are normally metabolized such as phenol, alcohols, aldehydes, ketones, or organic acids may exceed the oxidizing capacity of the resident microorganisms and cause sludge to disintegrate. The organisms are indifferent to concentrations of 1 ppm copper, 5 ppm of zinc, 2 ppm of nickel, or 10 ppm of chromium in influent waste.

All of the foregoing causes of sludge dispersal prevent acceptable settling, and produce a turbid effluent containing large numbers of bacteria, along with unoxidized waste. In wastewater treatment sludge that settles slowly without compacting, leaving only a small volume of supernatant liquid is called bulking sludge. Two types are distinguished: Filamentous and nonfilamentous bulking. In nonfilamentous bulking the particles of floc contain an excessive amount of water. This phenomenon, which depends upon the capacity of certain microorganisms to bind water, has been attributed to nutritional preferences among various genera of bacteria, in particular, the successful competition of *Zoogloea* in metabolizing specific kinds of waste. Pipes[2] cites as an example the apparent preference of *Zoogloea* for proteins and of *Sphaerotilus natans* for carbohydrates. Although these preferences are not conclusive, it can be inferred that *Zoogloea*, a water-binding genus, would predominate in a proteinaceous medium and *Sphaerotilus natans*, a filamentous species, would grow profusely on a carbohydrate waste. The former situation could lead to nonfilamentous bulking, and the latter to filamentous bulking.

In filamentous bulking numerous fungal or bacterial filaments extend from the individual particles of floc. This type of defect is characterized by profuse growth of filamentous bacteria, notably *Sphaerotilus natans*, *Bacillus macroides*, *Leucothrix*, and possibly *Beggiatoa*; by the fungus *Geotrichum candidum*; and probably by various yeasts such as *Rhodotorula*, *Torulopsis*, and *Trichosporon*. Other *Fungi Imperfecti* have also been implicated. Filamentous bulking is more likely in industrial waste than in domestic sewage treatment; it is especially troublesome in effluents from breweries, distilleries, sugar refining, and dairies. Soluble, easily metabolized chemicals, notably starch, sugars, albumin, and milk wastes favor the growth of filamentous bacteria, although high concentrations of any organic material can initiate filamentous bulking, as do sulfide, free sulfur, and deficiencies of nitrogen and phosphate. Coagulating polyelectrolytes improves the settling of sludge in some instances, but the additional expense may be prohibitive.

Another undesirable development in activated sludge treatment is a phenomenon called rising sludge, in which sludge settles well in the usual 30-min test, but then upon longer standing the entire compacted mass rises to the surface of the supernatant water in the graduated cylinder. There are two causes of rising sludge. The most common is the process of nitrification in the aeration vessel, followed by nitrate reduction in the sedimentation basin.

Aeration vessel: $$HNO_2 + 1/2\ O_2 = HNO_3 \tag{4.1}$$

Sedimentation basin: $$2HNO_3 = N_2 + H_2O + 5/2\ O_2 \tag{4.2}$$

The reaction in Eq. (4.1) is readily carried out by the nitrifying bacteria *Nitrobacter* and *Nitrocystis*. Only a few organisms can anaerobically reduce nitrate to nitrogen or nitrous oxide, but they appear now and then in waste treatment plants where the accumulation of gas bubbles from the reaction in Eq. (4.2) reduces the density of compacted sludge, causing the mass to rise. Anaerobic sludge also rises because of the entrapped bubbles of carbon dioxide, hydrogen, and hydrogen sulfide. This takes some time, however, and is usually the result of insufficient aeration or of stagnation in the sedimentation basin.

2. *THE ROTATING BIOLOGICAL CONTACTOR*

The rotating biological contactor is finding increasing industrial application as a method of secondary waste treatment, primarily because of its alleged simplicity and trouble-free operation. In principle, the process appeals strongly to the imagination, for in a single treatment vessel means are provided for aeration, agitation, microbiological growth, intimate contact between microorganisms and waste, and continuous self-stripping of the biomass. As with other biological oxidations, however, the process is adversely affected by low temperature, pH swings, shock loading, and toxicants.

The contactor intself consists of a series of large concentrically corrugated plastic discs fixed on a rotating shaft, which is mounted in bearings over a semicircular concrete basin through which waste water flows. When in operation, the shaft and the discs attached to it are rotated slowly in the vertical plane by a motor. Approxmately 40 percent of the lower portion of the discs are submerged at any moment. Then, as the discs rotate a film of waste water is carried into the air that flows downward over the microbiological film that develops on each disc. Oxygen is absorbed in the process allowing organic matter suspended or dissolved in the film of water to be continuously metabolized by the microbiological community attached to each disc. Initially, organisms naturally present in waste attach themselves to the discs, building a fixed layer a few millimeters thick containing 5–10 percent of solids. The film growth takes 18–24 h on the first few discs, but because of the progressive decrease in concentration of BOD through the unit, 5–7 days are required to film discs in the later stages.

According to Antonie,[5] four or more stages of discs provide the best removal of BOD and suspended solids. This arrangement achieves plug flow and makes the most efficient use of the available surface area. Staging can be accomplished by installing a series of three bulkheads between successive sets of discs across a single basin; in larger installations four (or more) complete units are used in series. Waste water flows through the contactor in a direction parallel to the shaft at a hydraulic loading of 1–

1.5 gal per day/ft^2 of disc area. Units have been described with discs 12 ft in diameter mounted on a single shaft 25 ft long, providing 50,000-60,000 ft^2 of surface. The units are covered to prevent the biomass from freezing in cold weather, and to protect the relatively fragile discs from wind damage.

Rotation of the discs causes the attached biomass to grow in rough filamentous tufts projecting into the wastewater, which thus exerts a pull on them. When the hydraulic shearing force exceeds the shear strength of the biomass, large aggregates of the excess are pulled off into the mixed liquor. In ideal operation the growth rate equals the rate of stripping. Solids are kept in suspension by the agitation of the discs, which have an optimum peripheral velocity of about 60 ft/min until the flow carries them into the clarifier where they settle rapidly. The concentration of suspended solids in the mixed liquor is usually about one-half that of the BOD in the influent–50-200 ppm in domestic wastes. The settled sludge contains 3-4 percent of solids, of which some 80 percent is volatile matter. The separated solids must be disposed of in drying ponds or land fill; there is no advantage to recirculating sludge in this process.

Both dissolved and suspended organic material are oxidized in the rotating biological contactor. In a pilot plant study, Torpey, et al.[(6)] found that BOD was reduced from 124 to 9 ppm in a ten-stage unit; suspended solids were decreased from 107 ppm in the influent to 9 ppm in the effluent.

In contrast to the activated sludge process, wherein a heterogeneous community of microorganisms share an intimate coexistence analogous to a grossly contaminated culture, the discs of a rotating contactor sustain successive cultures of biochemically specialized microorganisms that change progressively through the stages of the unit. Proceeding through the series of discs, BOD decreases continuously, and the fixed cultures alter progressively, becoming less filamentous as the microbiological films grow thinner. When the concentration of BOD falls below 30 ppm, nitrifying bacteria are able to compete with the more rapidly growing heterotrophic bacteria that metabolize carbon compounds. As a consequence, the concentration of ammonia begins to fall, while that of nitrite begins to rise as the reaction in Eq. (4.1) comes into play. At the same time, bicarbonate alkalinity and pH decrease because of the formation of nitric acid. The removal of ammonia is desirable for it imposes an oxygen demand on receiving waters, is toxic to fish when un-ionized,[(7)] is corrosive to copper and its alloys, consumes excessive amounts of chlorine, and facilitates the growth of algae.

Optimally, a rotating biological contactor can remove 90–95 percent of BOD, 95 percent of suspended solids, 20 percent of phosphate, 80 percent of hexavalent chromium, 100 percent of cyanide, and converts all ammonia to nitrate. As in all biological oxidations, efficiency declines rapidly below 55 F, and adequate amounts of nitrogen and phosphate must be available to sustain microbiological growth. A scheme has also been investigated(6) to remove residual nutrients by exposing the effluent from a rotating unit to algae, with a final purification through activated carbon to eliminate any remaining nondegradable organic matter.

In the first stages of a biological contactor the discs are covered with a dark grayish-brown film of zoogloeal and filamentous bacteria. Common species include *Zoogloea remigera* and *Sphaerotilus natans.* Toward the middle stages the bacterial films become thinner, and their color changes to a reddish-brown or coppery hue as genera of nitrifying bacteria such as *Nitrobacter* and *Nitrocystis* begin to proliferate. At low hydraulic loading the culture of nitrifiers moves closer to the first stage. The preponderance of predators in the final stages severely limits the survival of bacteria so that growth is sparse on the last discs in the unit. Here, among others, are seen *Vorticella*, a stalked ciliate, *Difflugia*, an amoeba, *Carchesium*, another protozoan, and *Philodina*, a rotifer. In a specially illuminated algal unit described by Torpey,(6) he identified *Nitzchia*, a diatom, *Oscillatoria*, a filamentous blue-green alga, *Arcella*, an amoeba, and *Phormidium*, another blue-green alga.

It might be inferred from the foregoing description that the rotating biological contactor is a trouble-free system for secondary treatment of waste. This is not the case, however, for despite claims to the contrary these units, with their plug-type flow, are sensitive to shocks and mishaps just as are all other methods of waste treatment. Changes in BOD load (e.g., doubling or halving the concentration of organic material in the influent) can drastically reduce the efficiency of the process. A variation of two pH units over as short a period as 2–5 h can also seriously inhibit biological oxidation. Temperature changes have significant effects on the growth rates of microorganisms, and a suddenly toxic influent (unfortunately not at all predictable) may quickly strip the microbiological growth from the discs. Furthermore, biological oxidation is dependent upon the free access of oxygen, which is often limited in these units by the elaboration of microbiological films as much as 4 mm thick. It is doubtful that oxygen can diffuse through a bacterial film thicker than 150 μ. There is also a good deal of crowding of the microbiological community in these films, which impedes the transfer of substrates, metabolites, waste products and gases putting the microorganisms at a considerable disadvantage in their struggle to survive. Finally, if the concentration of suspended solids is too high in the influent, discs become mechanically bridged and anaerobic.

In the preceding discussion the ideal operation of a rotating biological contactor has been described, together with estimates of its capabilities under the most favorable circumstances. It may now be of interest to consider an actual four-stage installation intended to reduce phenol in the effluent of an oxidation pond in order to comply with a local regulation. The system consists of four units in series, each containing 200 discs mounted on a single shaft with spacing of 1 1/8 in., center-to-center. Each of the four units is 20 ft long, 16 ft wide, and 13 ft high. Other pertinent operating information is summarized in Table 4.2.

TABLE 4.2

Operating Variables in a Four-Stage Biological Contactor

Diameter of discs (ft)	11
Area/disc (2 sides) (ft^2)	190
Discs/unit	200
Area/unit (ft^2)	38,000
Area/4 units (ft^2)	152,000
Flow rate (gpd)	648,000–854,000
Hydraulic loading ($gal/ft^2/day$)	4.3–5.7

As shown in Table 4.2, the hydraulic loading is three to four times the rate normally recommended. Consequently, poor efficiency with respect to biological oxidation would be anticipated. Table 4.3 shows average reductions of 25 percent BOD, 22 percent oil, and 60 percent phenols. Note that bacteria counts are higher in the effluent than in the influent, indicating that hydraulic loading is too great to allow predators to become established. Although suspended solids were not measured, these values also are undoubtedly increased. This operation is an example of the expensive methods often required to meet water pollution standards; the treating system is operated the year-round, but is necessary only in the winter when an associated oxidation pond is less efficient in oxidizing phenols.

TABLE 4.3

Performance of a Four-Stage Rotating Biological Contactor

		ppm				
Day	*Sample*	*BOD*	*Oil*	*Phenols*	*Dissolved oxygen*	*Bacteria*/ml
1	In	19	13	0.03	1.3	810,000
	Out	14	10	0.01	3.6	910,000
2	In	24	12	0.06	1.6	350,000
	Out	18	11	0.03	3.7	473,000
3	In	22	18	0.04	1.4	680,000
	Out	16	12	0.02	3.2	1,500,000
4	In	26	16	0.03	1.5	890,000
	Out	20	12	0.01	3.5	1,200,000

3. *WASTE STABILIZATION PONDS*

Of the three methods of secondary waste treatment discussed in this chapter, the biological oxidation pond is the simplest to build and operate, and if land is available it is also the most economical. All three treatments rely on the same biochemical principles, although the microbiology and method of transferring oxygen differs in each. The successful functioning of an oxidation pond depends upon a symbiotic relationship between algae and bacteria, in which algae produce oxygen for bacterial respiration, and bacteria, in metabolizing carbonaceous waste, generate carbon dioxide, which in turn is consumed by the algae. An open pond suffers the disadvantage of being at the mercy of the elements. Low temperatures and lack of sunlight severely impair the efficiency of natural aeration and biological oxidation, but if the retention time is ample—ideally two to four weeks—efficiency ranges from 70-90 percent as measured by the determination of BOD. This method of treating industrial waste has the great advantage of being much less sensitive to shock loading or inundations of storm run-off than are the other secondary treatments.

Ponds can be operated either aerobically or anaerobically. In the former mode carbon dioxide and water are the end products, in the latter organic acids and hydrogen sulfide are formed. Anaerobic operation permits greater loadings (up to 500 lb per acre per day) but with some loss of efficiency and an increase in nuisance odors. The anaerobic mode is often chosen for oxidizing wastes from meat and poultry processing, leather tanning, and food processing, all of which produce copious amounts of biodegradable organic matter. Wastes from paper processing, wineries, textile manufacturing, sugar refining, rendering, chemical processing, and petroleum refining are most often treated aerobically at moderate loadings of 20-30 lb per acre per day.

Oxidation ponds[8] are large impounding basins 2-20 ft deep (optimally 3-4 ft for aerobic operation) to which waste water is added at one point, and from which effluent is withdrawn at some other point selected to allow suitable residence time. The inlet pipe should discharge wastewater near the center of the pond so that incoming wastes can be mixed quickly, and distributed uniformly throughout the pond.[9] As a rule, the distribution of nutrients and vertical mixing are accomplished by the wind, although this element has negligible influence on the concentration of dissolved oxygen in the pond. On the other hand, the depth of the pond has much to do with the rate of production of oxygen, which is directly proportional to the penetration of sunlight. Effective illumination in most instances is limited to 7-12 in. below the surface; the rate of oxygen production at the surface may be five times that at a depth of 2-3 ft.

As in the other biological oxidations already discussed, settleable, suspended, and colloidal solids coagulate and collect on the bottom of oxidation ponds where they decompose and release nutrients that diffuse upward into the water. The bottom layers of a pond are anaerobic, and now and then release hydrogen sulfide if the pond becomes overloaded with organic material. The efficiency of aerobic oxidation decreases at night and during the winter when the sun is obscured by overcast and the temperature is low. The rate of production in summer is six to seven times that in winter when the growth of algae is inhibited by lack of light and by cold weather. The BOD of the effluent rises at those times when algae are demanding oxygen in competition with the bacteria. For the same reason pH falls during the winter and the concentration of ammonia rises to about 80 percent of the concentration of total nitrogen as algae consume oxygen rather than carbon dioxide. Less than 10 percent of ammonia nitrogen is removed in cold weather, in contrast to 80-90 percent in summer. A similar reduction of phosphate occurs in summer, but none is removed in winter. Loadings up to 100 lb per acre per day can be metabolized aerobically if mechanical aerators are installed in the pond, or if a series of ponds is available. In common with other methods of treating wastes, the operation of oxidation ponds is essentially empirical, although Oswald, et al.[10] have devised a method for computing the photosynthetic efficiency necessary to meet the oxygen requirements of a waste under various conditions of pond depth, retention time, and illumination.

The efficiency of an aerobic waste stabilization pond is vitally dependent on the health and activity of photosynthetic phytoplankton, for they are the principal source of oxygen required for bacterial respiration. Should these algae die, the pond immediately becomes septic and remains so until a new community of algae is established. Total counts of phytoplankton range from 1 million–5 million/ml in aerobic oxidation ponds. On a dry basis, they burgeon at an average rate of 30–35 tons per acre per year; the rate of growth during summer months may rise to 60–75 tons per acre per year. Rapidly growing unicellular algae (particularly *Chlorella* and *Scenedemus)* are desirable genera. Loadings greater than 60 lb of BOD per acre per day produce dense mats of the blue-green genera *Oscillatoria* and *Phormidium*, both of which infest polluted waters and generate objectionable odors resembling those emanating from pigpens. The cells of these genera contain gas so they float on the surface, where they contribute very little oxygen to the water. *Chlamydomonas* and *Euglena*, both unicellular flagellates, are also common flora in open ponds. When growing abundantly, these two genera have a fishy odor; moderate amounts are not objectionable. *Spirogyra*, *Ulothrix*, and *Anabena* are also often present.

Biological oxidation in ponds is carried out by bacteria of the *Pseudomonas*, *Flavobacterium*, *Achromobacter*, *Aerobacter*, and *Alcaligenes* genera; fecal streptococci and fecal coliform bacteria, if present, are eliminated rapidly by nutritional competition from the more robust types. Porges and MacKenthun[9] report that 100 species of fungi including nine *Fusarium* have been identified in oxidation ponds, as well as some 50 species of yeasts. The latter are most often seen in systems treating domestic waste. Protozoans, ciliates, and rotifers help to produce a clear effluent by preying on bacteria and unicellular algae. In addition, *Daphnia* and other small crustaceans thrive in stabilized aerobic systems where they use bacteria, algae, and fungi for food, thereby assisting in the clarification of the effluent. Nematodes (round worms) are more numerous in oxidation ponds than in other secondary waste treatments, especially in ponds with high concentrations of solids.

A common phenomenon in anaerobic, or incipiently anaerobic, lagoons is the development of a pink, red, or purple bloom of sulfur bacteria belonging to the family *Thiorhodaceae*. Cooper[11] identified two genera in industrial oxidation ponds as *Thiocystis*, which he found at a density of 4×10^8/ml, and *Chromatium*. Slettin and Singer[12] identified *Thiopedia rosea*, a unicellular, spherical organism 1-2 μ in diameter, growing in flat sheets embedded in capsular material. All of these are photosynthetic purple sulfur bacteria found only in water containing hydrogen sulfide and exposed to light. They oxidize hydrogen sulfide to sulfur, which is deposited within the cell.

$$CO_2 + 2H_2S \overset{light}{=} (CH_2O) + 2S + H_2O \tag{4.3}$$

In aging cultures the oxidation proceeds all the way to sulfuric acid.

$$2H_2O + 2CO_2 + H_2S \overset{light}{=} 2(CH_2O) + H_2SO_4 \tag{4.4}$$

In common with algae, purple sulfur bacteria are photoautotrophic. The above reactions do not take place in the absence of light, so they are obligate phototrophs as well as obligate anaerobes. As they are not competitive with algae they do not affect the density of the latter. The color of the sulfur bacteria is attributed to an intracellular pigmentation of red carotenoids that masks the green bacteriochlorophylls. By consuming hydrogen sulfide the purple sulfur bacteria function incidentally as biological deodorizers.

McKinney[13] points out that the bacterial community in a pond is composed of two distinct strata. The first is a mixture of bacteria that maintain their existence by metabolizing organic compounds in the waste stream. A secondary group, which predominates in most oxidation ponds, lives by metabolizing the lysed products of the first group. Autoflocculation of bacteria is essential to achieve a clear effluent; this does not occur among bacteria that are multiplying and highly motile. Fungi consume a large variety of organic compounds, but as they are filamentous they do not form compact settleable floc. At low oxygen tension, low pH, and low concentrations of nitrogen, fungi predominate, producing a turbid effluent. The bacterial population grows best in the pH range of six to nine, so it is adversely affected by the presence of sugars, alcohols, aldehydes, and ketones, all of which are oxidized to acids. Sulfonates, which are products of petroleum refining, block the reaction of carboxylase and hinder metabolism; bacteria adapt to it in time, however.

4. THE BIODEGRADABILITY OF ORGANIC MICROBICIDES

If blowdown water from recirculating cooling systems is processed by a secondary waste treatment, account must be taken of the possible effects upon microorganisms in activated sludge of biocides used to control microbiological growth in cooling water. Further, if the secondary effluent water is discharged to lakes, rivers, or oceans provision must be made for the safety of fish and other marine life. The toxicity of polluted water to fish is measured by a standard procedure,[14] with results expressed as Median Tolerance Limit, TL_m. This is the concentration at which 50 percent of a group of test fish can survive for a specified time of exposure, usually 96 h. Reactive volatile chemicals such as acrolein and chlorine seldom present any difficulty, for they quickly dissipate in the cooling system itself. Should ammonia or phenol be present, however, they chlorinate readily to chloramines and chlorophenols, respectively, which are highly toxic to most fish.

In Table 4.4 toxicity limits are shown for 90 percent survival of three-spined sticklebacks after 96 h of exposure to a number of organic compounds and proprietary formulations used or proposed for controlling microbiological growth in cooling systems. For example, the table indicates that nine out of ten sticklebacks, on the average, can survive for 96 h in the presence of a maximum of 3.2 ppm of a formulation containing 10 percent of methylenebisthiocyanate. Several of the microbicides listed in the table are discussed in detail in later chapters.

Most microbicides can be oxidized or biodegraded by the active organisms in secondary waste treatments provided the concentration of toxicants in the waste water is not so high that the treating system suffers toxic shock. In addition to foaming, detergents and surfactants of various kinds reduce the rate at which atmospheric oxygen dissolves in water, unfavorably altering the oxygen balance. The ease with which alkylbenzene sulfonates are biologically oxidized depends upon the configuration of the alkyl group. Those with linear side chains oxidize more easily than those synthesized from propylene polymers, which have branched side chains.

Resistance to biological degradation is shown especially by compounds containing a quaternary carbon atom in the chain most remote from the benzene ring. Other examples of molecular recalcitrance are ethylenediaminetetraacetic acid, triethanolamine, melamine, morpholine, cresols, thiocarbamates, dichlorophenols, trichlorophenols, and pentachlorophenol.

Residues of the latter may persist in soil for five years or more. Triazine derivatives used as herbicides are also highly persistant, as are the pesticides chlordane, DDT, dieldrin, and some others.

TABLE 4.4

Toxicity of Selected Microbicides to Fish

Active (%)	*Organic Compound*	$TL_{90\text{-}96}$ (ppm)
50	*N*-dodecylbenzyl-*N*, *N*, *N*-triethylammonium chloride	19
50	1,3-dichloro-5,5-dimethylhydantoin	10
10	Chloroethylenebisthiocyanate	5
10	Methylenebisthiocyanate	3.2
21 6	3, 5-dimethyltetrahydro-2H-1, 3, 5-thiadiazine-2-thione 2-mercaptobenzothiazole	2.5
20 10	1-alkylamino-3-aminopropane monoacetate 2, 4, 5-trichlorophenol	1.1
35	Dodecylguanidine hydrochloride	0.8
26	Pentachlorophenol	0.5
34 9 6 2	Sodium pentachlorophenate 2, 4, 5-trichlorophenate Other chlorophenates Sodium dimethyldithiocarbamate	0.2
20 5 5	Methyldodecylbenzyltrimethylammonium chloride Methyldodecylxylylenebis (trimethylammonium chloride) Bis(tri-*n*-butyltin)oxide	0.2
24	*N*-alkyldimethylbenzylammonium chloride	0.1

If biological oxidations were complete all organic compounds would become inorganic. That is, carbon, hydrogen, nitrogen, sulfur, and phosphorus would be oxidized completely to carbon dioxide, water, nitrate, sulfate, and phosphate. A very large number of organic compounds can be transformed in this way by the microorganisms residing in the upper layers of soil. Resistant compounds, however, especially pesticides and chlorophenols, percolate downward through the soil intact, and often eventually contaminate ground water.

Alexander[15] points out that a systematic approach to molecular recalcitrance and microbial fallibility as related to the biodegradability of microbicides is not possible because as yet no guiding principles have been formulated. Nevertheless, a few investigations have been made, and it has been established that organic functional groups become increasingly difficult to attack in the following order: carboxyl, aldehydic carbonyl, hydroxyl and amino, ketonic carbonyl, and methyl groups. A diversity of complex chemical reactions are also brought about by microorganisms, among them oxidation, reduction, esterification, hydrolysis, phosphorylation, demethylation, deamination, and dehydrogenation.[16]

Goldberg and Wershaw[17] found that two species of *Achromobacter* and *Flavobacterium* were able to use dodecylguanidine as their sole source of carbon, and also that the native microorganisms in a river mud could degrade approximately 5 percent of the biocide in 68 days. From an activated sludge tank, Stafford and Callely[18] isolated a heterotrophic bacterium resembling *Pseudomonas stutzeri* that was able to use thiocyanate as its sole source of cellular nitrogen and sulfur. Wehner and Hinz[19] subsequently suggested the possibility of a direct biodegradation of the organic thiocyanates by this bacterium. In studying the four quaternary ammonium salts didecyldimethylammonium chloride, dioctyldimethylammonium chloride, alkyl (C_{12}-C_{18})dimethylbenzylammonium chloride, and alkyl (C_{16}-C_{18})dimethylethylbenzylammonium chloride, Gawel and Huddleston[20] found that the aliphatic compounds were biodegraded faster than the aromatics by a mixed culture of microorganisms obtained from soil and raw sewage. At an initial concentration of 10 ppm all four quaternary ammonium salts were completely degraded within 48 h, regardless of the presence or absence of other organic material. At a concentration of 20 ppm, however, the rate of biodegradation was negligible. Data were also presented showing that an initial dose of 29 ppm of a quaternary ammonium salt in the recirculating water in an operating cooling system was reduced to less than 1 ppm within 3 h. This suggests that despite relatively high initial doses very low concentrations would subsequently be imposed on a waste treatment applied to combined effluent. Pentachloropehnol, on the other hand, was degraded only 20 percent in 10 days from an initial concentration of 10 ppm. Therefore, essentially all of it would find its way to and through a secondary treatment, and would ultimately be discharged unchanged. For this reason, the use of chlorinated phenolics as biocides is being rapidly curtailed. Raymond and Jamison[21] have summarized the chemical reactions of hydrocarbons brought about by the genus *Nocardia.* Several species oxidize hydrocarbons as their sole source of carbon for the biosynthesis of cells. These species grow rapidly on normal paraffins containing 5-28 carbon atoms, but any substitutions on the chain inhibit their growth. A few *Nocardia* degrade benzene to muconic acid or β-ketoadipic acid by cleaving the aromatic ring. Ethylbenzene, however, a product of catalytic reforming, is oxidized to phenylacetic acid with the aromatic ring intact.

The biodegradation of phenol is treated as a special topic in the following section. The very complicated transformation of lignin to humus and the degradation of cellulose by fungi are discussed in Chapters II and VIII.

5. BIOLOGICAL OXIDATION OF PHENOLS IN COOLING SYSTEMS

Phenolic compounds are objectionable in effluent water from petroleum refineries as they are toxic to aquatic life, they impose a high oxygen demand on receiving waters, and when chlorinated, they impart an offensive taste to the flesh of game-fish. Catalytic cracking of some of the heavier compounds in petroleum (perhaps resins) produces relatively small yields of phenol, a more abundant proportion of ortho-cresol, lesser amounts of meta- and para-cresols, several isomers of xylenol, and a few other phenolic derivatives.[22] Less significant sources of phenolics are caustic scrubbing of cracked gasolines, hydrocracking, and phenol treating in the manufacture of lubricating oils.

The reuse of phenolic process water for cooling was pioneered by the Sun Oil Company of Pennsylvania.[23,24,25] In this application water previously stripped of oil, sulfide, mercaptans, and suspended solids is returned to a cooling tower as make-up, where it is oxidized to carbon dioxide and other acids by several genera of bacteria. Sulfur compounds must be removed from the process water to avoid nuisance odors, corrosion within the cooling system, and interference with the desired microbiological activity. In a study of the operation of a pilot plant for oxidizing phenolic water, Lynn and Power[26] identified a number of bacteria in the bio-reactor, and determined their distribution. Five genera accounted for most of the microbial community: *Achromobacter*, *Pseudomonas*, *Micrococcus*, *Flavobacterium*, and *Bacillus.* Although no single genus predominated all of the time, *Pseudomonas* and *Achromobacter* were present in highest numbers most of the time. Bacterial species that are scavengers of phenol are aerobic, heterotrophic, and mesophilic; some are spherical, others are rod-shaped. According to Mohler and Clere,[25] the recirculating water in a cooling system operated for biological oxidation also contains unicellular algae, diatoms, nematodes and protozoans. Because of the low oxygen tension in these systems *Desulfovibrio desulfuricans* is likely to become troublesome. *Protozoa* do not survive at temperatures exceeding 36 C, so they are not normally seen in recirculating water in a typical cooling system.

The intense microbiological activity in the presence of phenols generates a substantial volume of flocculated microorganisms accompanied by a certain amount of adsorbed oil. The concentration of suspended solids in the recirculating cooling water is thus increased but, again according to Mohler and Clere,[25] as much as 110 ppm of floc did not impair heat trânsfer, so long as dispersants were used and heat exchangers were back-flushed daily.[27] They also found that auxiliary pH control was unnecessary because carbon dioxide (and probably other acids) generated in the biological oxidation was sufficient to keep the pH in the range of 7.4–8.0. Also, as the concentration of dissolved oxygen in the recirculating water was negligible, the corrosion rate of steel was confined to the range of 1.5–3.0 mpy, without the addition of corrosion inhibitors.

Biocides, of course, are not used in this mode of operation, but it is necessary to ensure an adequate supply of ammonia and phosphate in order to sustain the biological oxidation. Elkin et al.[23] found that without seeding or any other special procedures, phenol-oxidizing bacteria developed in the cooling water that reduced the average concentration of phenol from 26 ppm in the make-up water to 0.09 ppm in the blowdown. Furthermore, the greater part of the reduction was accomplished during the short time the water was falling through the tower. This invites the inference that a substantial proportion of the phenol is eliminated by direct air oxidation, for it is doubtful that bacteria alone could oxidize the compound this rapidly, even though in the presence of 80–100 ppm of phenol the density of bacteria may be as high as 5.0×10^8/ml. At 5–10 ppm of phenol plate counts average 3.0×10^6–5.0×10^6/ml. It is also possible that phenol is rapidly oxidized to an intermediate compound that does not react with 4-aminoantipyrine (the usual analytical reagent for determining phenol) after which further degradation proceeds more slowly.

Evans, et al.[28] have summarized the work done in elucidating the sequence of intermediate compounds formed in the oxidation of phenol, and have proposed a mechanism for one of the steps that remained doubtful, namely, the conversion of cis, cis-muconic acid to β-ketoadipic acid.

The complete oxidation proceeds through several intermediate compounds as follows:

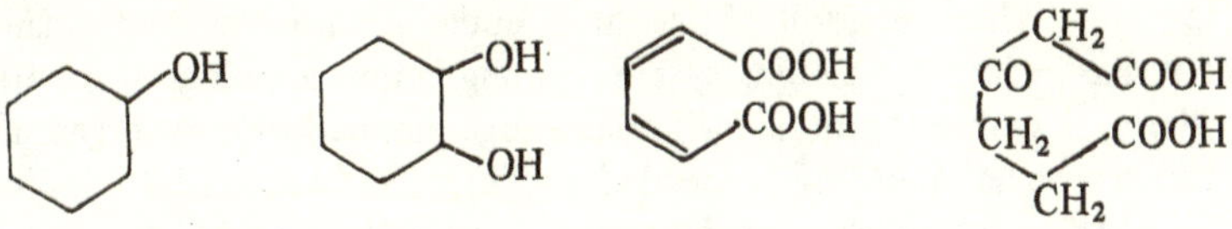

phenol *catechole* *cis,cis-muconic acid* *β-ketoadipic acid*

$$\begin{array}{l} CH_2COOH \\ | \\ CH_2COOH \end{array} \qquad CH_3COOH \qquad HCOOH \qquad H_2CO_3$$

succinic acid *acetic acid* *formic acid* *carbonic acid*

It is worth noting in passing that a few *Nocardia* degrade benzene to muconic acid or β-ketoadipic acid.[21]

6. *TREATED EFFLUENT AS MAKE-UP FOR COOLING SYSTEMS*

In the preceding section the special case of using a cooling tower for the biological oxidation of phenol in otherwise relatively clean process water was discussed. Some of the ramifications of using the effluent from secondary waste treatment as make-up for a recirculating cooling system are considered next. Note that this application differs from that in which phenol is intentionally oxidized. Here it is assumed that an effluent in which BOD, oil, and suspended solids have been reduced to a minimum—in effect, an effluent suitable for discharge to receiving waters, is used. This treated water will be used as make-up for a recirculating system as a conservation measure replacing fresh water. The blowdown water from the cooling system is discharged to the waste treatment plant and recycled. In general, easily oxidized organic compounds such as alcohols, phenols, and organic acids are absent. Upsets in the waste treatment process, however, give rise to a number of undesirable conditions in cooling systems using the effluent that demand consideration. If too much organic material (organic loading) is forced upon the treating plant, or if the rate of flow (hydraulic loading) is too great, the formation of activated sludge is impaired, and organic matter passes into the cooling system turning it into a biological reactor. Similarly, if sedimentation is incomplete large numbers of microorganisms and excessive amounts of suspended solids are introduced into cooling water. Then too, such microbiological contamination includes several types not normally encountered in cooling systems supplied with fresh water, among others the Gram-positive *Bacillus*, which forms spores highly resistant to biocides, *Zoogloea remigera*, a prolific slime-former, *Sphaerotilus natans*, a filamentous species of bacteria that forms sticky mats, along with an assortment of molds, yeasts, and fungi. In the presence of an abundant supply of carbon compounds it is impossible to control the growth of these microorganisms, consequently the recirculating water quickly teems with biological floc, and plugging of the cooling system ensues. These difficulties arise more often in effluents from rotating biological contactors and activated sludge plants than in those from waste stabilization ponds, so it is essential that the former processes receive scrupulous attention if the effluent is to be used successfully for cooling water.

Other components of effluent from secondary waste treatment that can cause complications in cooling water are detergents, ammonia, and phosphate. Some resistant detergents pass through waste processes unchanged, leading to foaming, both in the treating plant and in cooling water. Ammonia, which may be present at 5-15 ppm, or more, can in the presence of oxygen, lead to severe corrosion of copper alloys. Phosphate, usually present in treated effluents at 2 ppm, or higher, causes scales of calcium and iron phosphates in heat exchangers, and also stimulates algal growth in cooling towers. A further complication is that microorganisms in secondary effluents are more likely to be acclimated to microbicides than are those that enter a cooling system in wind-borne dirt, making control still more difficult.

Table 4.5 contains a list of salts and contaminants with their recommended limits of concentration in recirculating water. These values are not rigid requirements, but rather estimates of the maximum concentrations that can be tolerated without undue difficulty. Neither would it be likely that any single cooling water would be encountered in which all of the variables were at these maximum values. In modern industrial water management it is necessary to take what is available and make do.

TABLE 4.5

Limitations in Recirculating Cooling Water

Component	*Maximum Concentration* (ppm)
Silica, SiO_2	175
Chloride, Cl^-	400
Sulfate, SO_4^{--}	625–835
Calcium Hardness, $CaCO_3$	1500–2000
Total Dissolved Solids	3500
Suspended Solids	100–150
Biochemical Oxygen Demand (BOD)	50
Oil	25
Viable Plate Count	10^6 (bacterial/ml)

One by one the components listed in Table 4.5 are considered with respect to their significance in cooling water. The concentration of silica is limited to 175 ppm because of its tendency to form scales with magnesium. Also, it is essential to the reproduction of diatoms, whose cell walls are siliceous. Many industries located along seacoasts use sea water for once-through cooling, after which it may be combined with treated effluent for discharge. Where this is the practice, it is common to find halophilic bacteria(29) thriving in the combined effluent. These organisms are sensitive to low pH, and if the salt concentration is reduced by dilution they swell and rupture. At concentrations exceeding 400 ppm of chloride, or perhaps even less, stainless steels are likely to be pitted and are also subject to stress-corrosion cracking. Sulfate at excessive concentration is precipitated by calcium ions as calcium sulfate, anhydrite, or $CaSO_4 \cdot 2H_2O$, gypsum. There is a rule of thumb (not a very useful one, incidentally) that to avoid calcium sulfate scaling the concentration of calcium multiplied by that of sulfate, both in parts per million, should not exceed 500,000. In addition to precipitating as calcium sulfate, high concentrations of calcium (and other positive ions) decrease the bactericidal activity of certain anionic surfactants,(30) and by competing for negative sites on the cell walls of bacteria, they also diminish the activity of quaternary ammonium salts.(31) High concentrations of both chloride and sulfate increase the corrosiveness of cooling water. Both aerobic and anaerobic treatments generate an effluent containing substantial numbers of the sulfate-reducing bacterium

Desulfovibrio desulfuricans, which thrives in the presence of sulfate. Toerien et al.[32] counted 30,000-50,000 sulfate reducers per milliliter in anaerobic activated sludge digesters; the numbers of organisms increased with additional sulfate. Also, hydrogen sulfide forms rapidly in effluents from aerobic processes containing a high concentration of sulfate, showing that sulfate-reducing bacteria are on hand in large numbers. The gradual precipitation and settling of organic matter in areas of low flow within a cooling system results in the accumulation of nutrients suitable for the growth of a variety of microorganisms. The depletion of oxygen and the generation of carbon dioxide by these organisms favors the growth of sulfate-reducing bacteria, provided a source of sulfate is available. Unfortunately, there is no way to assess the growth of microorganisms in stagnant passages in a cooling system. High concentrations of dissolved solids increase corrosion rates, scaling, and the tendency of cooling water to form deposits. In addition, microbiological control is made more difficult by the presence of large amounts of dissolved material. Suspended solids and BOD bring together microorganisms and nutrients, leading upon settling to profuse biological growth with under-deposit corrosion. The accumulation of this material in the basin of a cooling tower generates odors, reduces chromate, and precipitates iron and zinc as a gelatinous black slime. Hydrocarbons in cooling water coat cooling tower fill as well as heat transfer equipment; they also interfere with biological oxidation of some compounds, but if emulsified they become accessible to bacteria permitting the latter to increase in number.

In an arrangement where the blowdown from a cooling system is returned to the waste treatment plant for recycling, it is essential that indiscriminate addition of biocides to cooling water be avoided, otherwise the secondary waste treatment process may suffer toxic shock leading to an increase in suspended solids, objectionable algal turbidity in oxidation ponds, or to an excessive number of microorganisms in the effluent. In this type of mishap protozoans and rotifers, which are more vulnerable to biocides than bacteria, are killed producing a cloudy effluent containing numerous bacteria. In this situation the impulse is to add more biocide to the cooling system, which of course, only exacerbates the problem. Furthermore, it is unlikely that a secondary infestation of protozoans could arise within the cooling system to reduce the numbers of bacteria in the cooling water, as these higher organisms do not tolerate warm water. Thus, it is essential that the biological oxidation process be under control at all times if its effluent is to be used for cooling.

7. *CONCENTRATION RELATIONSHIPS IN RECYCLED BLOWDOWN*

It is of interest to calculate the increase in the concentration of solids in a secondary waste treating system such as described in preceding sections, in which the blowdown water from cooling systems is recycled. Fig. 4.1 illustrates the flow of water through a waste treatment plant, represented in the diagram by a rectangle. The make-up to the system is process water of some kind, part of which is used as make-up for cooling systems with the surplus discharged to receiving waters.

In the process represented by Fig. 4.1 let

V_m = make-up to system (gpd)
V_s = flow to receiving waters (gpd)
V_e = rate of evaporation from cooling towers (gpd)
V_r = flow of returned blowdown from cooling systems (gpd)
V_T = total flow in system (gpd)
c_s = concentration of solids in system (lb/gal)
c_m = concentration of solids in make-up (lb/gal)

Also,

$$V_T = V_m + V_r$$

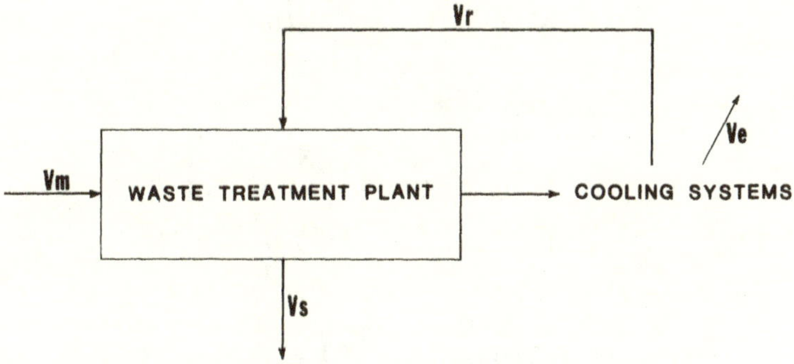

Fig. 4.1 Flow of water through a waste treatment plant.

If a material balance of the solids entering and leaving the treating plant is made, it is apparent that whatever goes on in the cooling system loop, the total solids in the system is governed by the relationship

$$Vm = V_s + V_e$$

Also

$$\text{Total solids} = V_m c_m - V_s c_s$$

The rate of change of solids in the system is expressed by the following differential equation:

$$dc_s/dt = (V_m c_m/V_T) - (V_s c_s/V_T) \qquad (4.5)$$

This equation can be solved by separating variables and integrating.

$$\frac{dc_s}{(V_m c_m/V_T) - (V_s c_s/V_T)} = dt$$

Also, to simplify the notation, let

$$\alpha = V_m c_m/V_T \qquad \text{and} \qquad \beta = V_s/V_T$$

Then,

$$\int_{c_0}^{c_s} dc/(\alpha - \beta c) = \int_{t_0}^{t} dt$$

$$(-1/\beta)\log_e(\alpha - \beta c)\Big]_{c_0}^{c_s} = t\Big]_{t_0}^{t}$$

$$(-1/\beta)\log_e\left(\frac{\alpha - \beta c_s}{\alpha - \beta c_0}\right) = (t - t_0)$$

Or, solving for c_s

$$(\alpha - \beta c_s) = (\alpha - \beta c_0)e^{-\beta(t-t_0)}$$

$$c_s = \alpha/\beta + (c_0 - \alpha/\beta)e^{-\beta(t-t_0)}$$

And finally

$$c_s = \frac{V_m c_m}{V_s} + \left(c_0 - \frac{V_m c_m}{V_s}\right) e^{-(V_s/V_T)(t - t_0)} \qquad (4.6)$$

As t (in days) becomes large the exponential factor becomes small, and in a few days the right-hand term vanishes. Therefore, the maximum value of c_s is $V_m c_m/V_s$.

References

(1) Parker, H. W. 1975. *Wastewater systems engineering.* Englewood Cliffs: Prentice-Hall.

(2) Pipes, W. O. 1966. "The Ecological Approach to the Study of Activated Sludge," In *Advances in applied microbiology,* vol. 8, W. W. Umbreit, ed., New York and London: Academic Press.

(3) Painter, H. A., Viney, M., and Bywaters, A. 1961. Composition of sewage and sewage effluents. Inst. Sewage Purif., *J. Proc.* 1961:302.

(4) Pipes, W. O. 1967. Bulking of activated sludge. In *Advances in applied microbiology,* vol. 9, ed. W. W. Umbreit, New York and London: Academic Press.

(5) Antonie, R. L. 1976. *Fixed biological surfaces–wastewater treatment.* Cleveland: CRC Press.

(6) Torpey, W. N., Heukelekian, H., Kaplovsky, A. J., and Epstein, R. 1971. Rotating disks with biological growths prepare wastewater for disposal or reuse. *J. Water Pollut. Control Fed.,* 43:2181.

(7) McCoy, J. W. 1974. *The chemical treatment of cooling water.* p. 204, New York: Chemical Publishing Company.

(8) McKinney, R. E. 1962. *Microbiology for sanitary engineers.* Chapter 22, Oxidation ponds. New York: McGraw-Hill.

(9) Porges, R., and MacKenthun, K. M. 1963. Waste stabilization ponds: use, function, and biota. *Biotechnology and Bioengineering,* 5:255.

(10) Oswald, W. J., Gotaas, H. B., Golueke, C. G., and Kellen, W. R. 1957. Algae in waste treatment. *Sewage Ind. Wastes* 29:437.

(11) Cooper, R. C. 1963. Photosynthetic bacteria in waste treatment. In *Developments in industrial microbiology,* vol. 4, Amer. Inst. Biol. Sci., Washington, D. C.

(12) Slettin, O. and Singer, R. H. 1971. Sulfur bacteria in red lagoons. *J. Water Pollut. Control Fed.* 43:2118.

(13) McKinney, R. E. 1957. Microbiological progress report. activity of microorganisms in organic waste disposal. II. aerobic processes. *Appl. Microbiol.* 5:167.

(14) Am. Soc. Testing Materials. 1977. Evaluating acute toxicity of water to fresh–water fishes. Annual Book of ASTM Standards, Part 31, D1345-59.

(15) Alexander, M. 1965. Biodegradation: problems of molecular recalcitrance and microbial fallibility. In *Advances in applied*

microbiology, vol. 7, ed. W. W. Umbreit, New York and London: Academic Press.

(16) Sebek, O. K. 1971. Microbial transformations of antibodies. In *Advances in applied microbiology,* vol. 14, ed. D. Perlman, New York and London: Academic Press.

(17) Goldberg, M. C. and Wershaw, R. L. Sept. 1965. *Biodegradation of dodecylguanidine acetate (iodine).* Abstract of Papers, p. 17x, Item 40, 150th Meeting of Am. Chem. Soc., Atlantic City, N.J.

(18) Stafford, D. A. and Callely, A. G. 1969. The utilization of potassium thiocyanate by a heterotrophic bacterium. *J. Gen. Microbiol.,* 55:285.

(19) Wehner, D. C. and Hinz, C. F. 1971. Organic thiocyanates as industrial antimicrobial agents. chapter 45, vol. 12, *Developments in industrial microbiology.* New York: Plenum.

(20) Gawel, L. J. and Huddleston, R. L. April 23–26, 1972. *The biodegradability of low concentrations of certain quaternary ammonium antimicrobials by bacteria.* Presented at the American Oil Chemists National Meeting, Los Angeles, California.

(21) Raymond, R. L. and Jamison, V. W. 1971. Biochemical activities of *Nocardia.* In *Advances in applied microbiology,* vol. 14, ed. D. Perlman, New York and London: Academic Press.

(22) Gruse, W. A. and Stevens, D. R. 1960. *Chemical technology of petroleum.* 3rd ed., p. 67, New York: McGraw-Hill.

(23) Elkin, H. F., Mohler, E. F., and Kumnick, L. R. 1956. "Biological Oxidation of Oil Refinery Wastes in Cooling Tower Systems," *Sewage Ind. Wastes,* 28:1475.

(24) Mohler, E. F., Jr., Elkin, H. F., and Kumnick, L. R. Nov. 1964. Experience with reuse and bio-oxidation of refinery wastewater in cooling tower systems. *J. Water Pollut. Control Fed.* 36:11, 1380.

(25) Mohler, E. F. and Clere, L. T. Oct., 1973. Bio-oxidation process saves H_2O. *Hydrocarbon Processing* 52:10, 84.

(26) Lynn, G. E. and Powers, T. J. Jan., 1955. Bacterial studies in oxidation of phenolic wastes. *Sewage Ind. Wastes* 27:1, 61.

(27) Hitzeman, T. E. and Feist, H. W. July, 1958. To clean your condenser–backwash is cheapest. *Power Eng.* 62:7, 72.

(28) Evans, W. C., Smith, B. S. W., Linstead, R. P., and Elvidge, J. A. 1951. Chemistry of the oxidative metabolism of certain aromatic compounds by micro-organisms. *Nature* 168:772.

(29) Kushner, D. J. 1968. Halophilic bacteria. In *Advances in applied microbiology,* vol. 10, eds. W. W. Umbreit and D. Perlman, New York and London: Academic Press.

(30) Voss, J. G. 1963. Effect of inorganic cations on bactericidal activity of anionic surfactants. *J. Bact.* 86:207.

(31) Chaplin, C. E. 1952. Bacterial resistance to quaternary ammonium disinfectants. *J. Bact.* 63:453.

(32) Tuerien, D. F., Thiel, P. G., and Hattingh, M. M. 1968. Enumeration, isolation, and identification of sulfate-reducing bacteria of anaerobic digestion. *Water Research* 2:7, 505.

V.

The Control of Microorganisms

In the first four chapters, the kinds of microorganisms that live in recirculating cooling water systems, their effects on these systems, and factors governing their health and growth are discussed. In order to limit the number of bacteria in cooling water, chemicals can be added to achieve either of two physiological effects. If inhibition of growth is desired bacteriostatic agents are used, which interfere with reproductive and metabolic processes. If the death of all microorganisms is the goal a microbicide is used. There is considerable ambiguity in these terms, however, as the effect obtained depends upon concentration of chemical, absorption rate, and a number of other variables. In general, microbicides are appropriate for cooling water, for although in principle it might be feasible to limit the number of bacteria to some presumably innocuous value by using suitable bacteriostatic agents, it is seldom readily apparent what that value should be. Also, allowing a constant population to exist encourages the development of resistant mutants or strains that can become exceedingly troublesome.

Bungay(1) cites six ways in which bacteria can adapt to a microbicide:

1. The organism produces an enzyme that destroys the biocide.
2. A change is brought about in the internal structure of the cell.
3. The permeability of the cytoplasmic membrane is altered to prevent the entry of the biocide.
4. A change takes place in the composition of the cell wall.
5. A spontaneous mutation occurs on the chromosome or in a plasmid.
6. An infectious transfer of resistance is accomplished.

Bacteria manufacture a diversity of enzymes and show a remarkable ability to metabolize toxic chemicals. Pseudomonads, for example, secrete enzymes that permit them to use as carbon sources thymol, cresols, phenols, nitrophenols, chlorophenols, quaternary ammonium salts, and a number of other normally toxic compounds. Some bacteria, most commonly *Bacillus* and *Clostridium*, when they find themselves in threatening surroundings, develop an endospore within the body of the cell. The spore thus formed is stubbornly resistant to heat, cold, and most chemicals. Some chemicals stimulate or inhibit permease, the enzyme that regulates the permeability of the cytoplasmic membrane; when the permeability decreases larger molecules cannot gain entry into the cell.

An alteration in the composition of the cell wall is an extremely effective method of self-protection by bacteria. Thus, Chaplin[(2)] found that a particular species of Gram-negative bacteria, *Serratia marcescens*, normally susceptible to 100 ppm of an alkyldimethylbenzylammonium chloride, could be adapted to grow in 100,000 ppm of the chemical. Staining with Sudan Black B, a fat-specific stain, showed that the adapted cells had elaborated a lipid or lipoprotein complex on the surface of the cell that was capable of resisting the disruptive forces arising from the surface activity of the quaternary ammonium salt. It was also verified that some members of the bacterial population were naturally more resistant than others, which was attributed to their possession of more lipoprotein or phospholipid than normal in their cell walls.

A small change in the genome of a bacterium can make the cell resistant to a certain type of microbicide. The genome is the totality of genetic information encoded in deoxyribonucleic acid, which controls the development, functioning, and reproduction of a cell. Kiser et al.[(3)] quote the probability of a single spontaneous mutation on the single bacterial chromosome conferring resistance to drugs on a certain species of bacteria as 3×10^{-6} to 6×10^{-10}. If this single mutation conferred resistance to one microbicide, then for the organism also to develop resistance to an unrelated microbicide a second mutation would have to occur at some other point on its chromosome. The probability of two simultaneous mutations conferring resistance to two unrelated biocides is extremely small. In fact, it is the product of the probabilities for single mutations, i.e., 9×10^{-12} to 3.6×10^{-19}. This suggests combining unrelated biocides, which no doubt would be an effective ploy against a single species. In practice, however, little is gained by using mixtures of biocides against the variety of microorganisms that are found in cooling water. The infectious transfer of resistance is accomplished by a process of conjugation whereby an extra-chromosomal element called the resistance transfer factor (a plasmid, or miniature chromosome) is implanted in a recipient bacterium by a donor by a mechanism analogous to sexual conjugation.

The effect of microbicides may be general or specific; they may be toxic to a large number of taxonomic groups, or to a limited number of species. The former are preferred, and are often described in promotional literature as having a "broad spectrum of biological activity." Combining several microbicides in a single formulation sometimes increases the killing efficiency: more often it doesn't. Regardless of the microbicide used, however, its first point of attack is upon the cell wall, which conversely is the first line of defense of a microorganism. Thus, older cells having well-developed extramural capsules, or slime layers, are significantly more resistant than younger rapidly respiring cells, which tend to absorb toxic chemicals at a higher rate. Also, bacteria exposed to sublethal concentrations of toxicants develop changes in their cell walls that make them more resistant to chemicals. Lowick and James[(4)] found, for example, that the cell wall of a strain of *Aerobacter aerogenes* that had become immune to crystal violet contained a surface layer of lipid instead of the normal polysaccharides. Indeed, the cell walls of most resistant bacteria are characterized by a high content of lipid.

The second defense of a microorganism is the cytoplasmic membrane—the osmotic barrier controlling the passage of metabolites into and out of the cell. Almost invariably this membrane contains the cytochrome pigments vital to respiration, as well as permeases that regulate the permeability of the living cell. Chemicals that inactivate these enzymes cause the death of the cell through paralysis of respiration, or by altering the cytoplasmic membrane so that leakage of metabolites ensues.

Death, like growth, follows a simple exponential rate law, which is influenced by three factors[(5)]:

1. The mass transfer of toxicant in solution to the bacteria-water interface.
2. The chemisorption of toxicant at active centers on the cell wall.
3. The diffusion of the activated chemisorbed complex with chemical attack on cellular or intracellular structures.

If the toxicant is in excess the decrease in numbers can be expressed by the equation:

$$\log_e (N_1/N_2) = k(t_2 - t_1) \tag{5.1}$$

where N_1 is the number of viable organisms at time, t_1, when the toxicant is added, and N_2 is the number remaining at a later time t_2; k is the rate constant. Sterility is never obtained in cooling systems because the death of microorganisms is exponential, with the further complication that an open recirculating cooling system is constantly being inoculated by windborne organisms.

1. *MICROBICIDES AND THEIR MODE OF ACTION*

Before a biocide can be marketed, it is necessary to accumulate laboratory data showing its effectiveness against various microorganisms. Ordinarily this is done by determining minimum inhibitory concentrations, examples of which are shown in Table 6.16 p. 142. These data are then submitted to the Environmental Protection Agency in support of an application for an Experimental Permit. After the permit is issued, the product must be tested in water systems in which it is proposed to be used for a period of not less than 30 days during two different seasons of the year. Reports of these tests must include, among other things, the concentration of biocide used, the interval between treatments, bacterial counts before and after treatments, and visual ratings of the type and extent of microbiological growth. Upon successful completion of this program, the product is issued a registration number by the Environmental Protection Agency. Some of the toxic chemicals used for keeping the population of nuisance organisms under control are considered next, separating them for convenience into three groups according to their toxic effects.

1.1 Oxidizing Agents

Oxidizing agents that have the widest application for microbiological control in water systems are the halogens: chlorine, bromine, and iodine. As chlorine is the most frequently used in industry, this discussion is confined to chlorine and chlorine compounds. Bromine and iodine are used to sterilize swimming pools and in a few other special applications, but their cost is too great for large-scale industrial use.

Chlorine

Chlorine is a poweful oxidizing agent that combines readily with protoplasm, forming stable nitrogen-chlorine bonds with proteins. It is, therefore, toxic to all living organisms, and at high concentrations (2-3 ppm) produces general devastation within cells. The effectiveness of chlorine is enhanced by low pH, high temperature, and high concentration. It is especially efficient for sterilizing potable waters, as a relatively high concentration can be maintained for long periods in the closed piping of municipal distribution systems. It is for this reason that few if any bacteria are introduced into cooling systems in make-up water supplied by a municipality. Most ground waters (wells, springs) also are relatively free of bacteria.

The effectiveness of chlorine is impaired by organic matter, including oil and masses of algae or slime; by reducing agents such as hydrogen sulfide and sulfur dioxide; and by chemicals containing reactive nitrogen, e.g., ammonia, polyacrylamide, and alkanolamines. Sufficient chlorine must be added to the water to satisfy all these reactions, which are collectively measured as "chlorine demand,"[(6)] and to allow a suitable concentration of residual chlorine to persist in the water. This process is called "Breakpoint chlorination."

Chlorine oxidizes sulfhydryl groups of certain enzymes, such as triose phosphate dehydrogenase, and inactivates reducing enzymes that are essential to respiration. Algae are easier to kill than bacteria, but when relatively thick masses are present, only those on its surface can be reached by dissolved chlorine. *Aerobacter* and *Pseudomonas* in neutral solution are destroyed within five minutes by a concentration less than 0.1 ppm of chlorine at 20-25 C. Unlike most other microbicides chlorine destroys spores at similarly low concentrations, but *Desulfovibrio desulfuricans* develops a resistance to chlorine.

When chlorine gas dissolves in water it undergoes a rapid reversible hydrolysis, and the pH falls significantly.

$$Cl_2 + H_2O = HClO + H^+ + Cl^- \qquad (5.2)$$

The toxicant in an aqueous solution of chlorine is the weak acid, hypochlorous acid, formed in the hydrolysis. In general, the cytoplasmic membrane is much more permeable to undissociated molecules than it is to ions; the ability to pass through this membrane is an important factor in determining toxicity. For this reason, hypochlorous acid is toxic to living cells, whereas hypochlorite ion is relatively innocuous.

Latimer[(7)] gives the value of the hydrolysis constant as $k_h = 4.66 \times 10^{-4}$; his value for the dissociation constant of hypochlorous acid is $k = 3.2 \times 10^{-8}$. As pH rises hypochlorous acid is neutralized, forming an increasing concentration of the inactive hypochlorite ion.

$$HClO + OH^- = H_2O + ClO^- \qquad (5.3)$$

Chambers and Clark[(8)] quote the values summarized in Table 5.1 for the percent of undissociated hypochlorous acid at various values of pH.

TABLE 5.1

Effect of pH on the Dissociation of Hypochlorous Acid

pH	HClO (%)
4	100
5	99.7
6	96.8
7	75.2

Both the hydrolysis of chlorine and the dissociation of hypochlorous acid are favored by increasing alkalinity, giving the net reaction represented by Eq. (5.4):

$$Cl_2 + 2OH^- = H_2O + Cl^- + ClO^- \tag{5.4}$$

Thus, the efficiency of chlorine as a microbicide is inversely proportional to pH; its toxicity is weak at pH 8 and negligible at pH 9.

Off-setting to some extent the low cost of chlorine treatment are several disadvantages that should be considered. The element is difficult and dangerous to handle, and it is extremely corrosive to many metals, as shown by the following general reaction.

$$nCl_2 + 2M = 2MCl_n \tag{5.5}$$

Maintenance of a chlorinator and associated piping is, therefore, costly. Reactions with a number of contaminants often encountered in cooling water produce hydrogen ions that lower the pH of the recirculating water.

$$Cl_2 + H_2S = S + 2H^+ + 2Cl^- \tag{5.6}$$

$$2H_2O + Cl_2 + SO_2 = SO_4^{--} + 4H^+ + 2Cl^- \tag{5.7}$$

$$Cl_2 + NH_3 = NH_2Cl + H^+ + Cl^- \tag{5.8}$$

It has been established that chlorine renders both pine and redwood susceptible to attack by fungi, apparently by solubilizing the naturally toxic extractives in the wood. It also reacts with lignins to form chlorolignins, which being soluble in slightly alkaline water, thus it destroys fill slats and mist eliminators by chemical deterioration. Also, chlorine reacts with a number of water-treating chemicals including ferrocyanide, polyacrylamide, and ligninsulfonates, interfering with their functions, and being itself inactivated as a biocide. Considerable chlorine is dissipated from the warm water as it rains down through the tower, often reducing its concentration in the basin to a negligible value. Finally, chlorine reacts slowly with water in sunlight, releasing oxygen, and further depleting the concentration of available chlorine.

$$Cl_2 + H_2O = 2H^+ + 2Cl^- + \frac{1}{2}O_2 \qquad (5.9)$$

Chlorine Compounds

Several *N*-chlorinated compounds that are sometimes used as sources of "active chlorine" are chloramine-*T*, or para-toluene sodium sulfonchloramide, the formula of which is $CH_3\text{-}C_6H_4\text{-}S{:}O\cdot(ONa){:}NCl\cdot H_2O$; *N*-chlorosuccinimide, $(CH_2CO)_2N\text{-}Cl$; 1, 3-dichloro-5, 5-dimethylhydantoin; and trichlorocyanuric acid. In general, the activity of the *N*-chlorine derivatives is much less than that of the free element, except above pH 9, and they are, of course, much more expensive. Petterson, and coworkers[9,10] have studied the hydrolysis and properties of one of these compounds, and found that no actual chlorine, and only transient amounts of hypochlorous acid are relased. Whatever the form of the active chlorine, however, it is iodometrically titratable and reacts characteristically with ortho-tolidine.

The synthesis of several chlorinated hydantoins has been described by Ishii, et al.[11] One that has been thoroughly investigated,[9] and is available commercially is 1,3-dichloro-5,5-dimethylhydantoin.

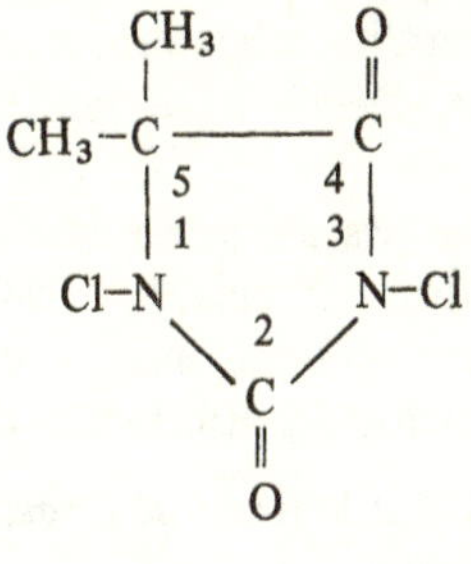

When this chemical is added to water nitrogen trichloride is evolved, and the pH falls about two units as the solid slowly decomposes. At the end of one hour 50–60 percent of the apparent active chlorine has disappeared. Other products of the hydrolysis are 1-chloro-5,5-dimethylhydantoin, *N*-chloroisopropylamine, chloride ion, nitrogen, and carbon dioxide. The presence of *N*-chloroisopropylamine, $(CH_3)_2CHNHCl$, shows that cleavage of the hydantoin ring occurs. Nitrogen trichloride would be expected to hydrolyze to hypochlorous acid:

$$2NCl_3 + 3H_2O = 3HClO + N_2 + 3H^+ + 3Cl^- \qquad (5.10)$$

but if HClO is ever present in the reaction mixture, it apparently exists only as an intermediate. The monochlorohydantoin also possesses chlorine-like activity.

A similar compound is trichlorocyanidine (also called tricyanogen chloride and cyanuric trichloride), which on hydrolysis gives cyanuric acid, nitrogen, chloride ion, and carbon dioxide; hypochlorous acid and nitrogen trichloride probably are intermediates. The proper name for this compound is trichloro-*s*-triazine-2,4,6-(1H,3H,5H)-trione.

Chlorine Dioxide

Chlorine dioxide is an unusually powerful microbicide that is more lethal than chlorine to both vegetative cells and spores. The dioxide is a gas that explodes with great violence at concentrations greater than 15 percent, so it is either generated *in situ* from sodium chlorite and chlorine at such a rate that it is immediately dispersed in the cooling water, or else it can be purchased stabilized in a relatively dilute carbonate solution.

$$ClO_2^- + \tfrac{1}{2}Cl_2 = ClO_2 + Cl^- \qquad (5.11)$$

Commercial formulations are now available* containing either 6 or 12 percent equivalent ClO_2. As yet this chemical has not found wide acceptance for cooling water systems, but when properly applied it is exceptionally efficient and economical. Details of its application are given in Chapter VI.

Stabilized chlorine dioxide has a number of advantages over chlorine. It does not chlorinate aromatics, ammonia, or unsaturates, so it is not inactivated or consumed by these contaminants. It does not volatilize from dilute solution, and it is not markedly affected by changes in pH, although

* Rio Linda Chemical Company, 1311 I Street, Rio Linda, California 95673

its toxicity is enhanced at higher pH, and reduced to some extent at low pH. This latter property is a distinct advantage in systems under nonchromate treatments, which in general, are alkaline. Finally, the dioxide is persistent, whereas chlorine is not; it is the oxidizing agent of choice in cooling waters having excessive chlorine demand.

The antibiological activity of chlorine dioxide lies entirely in its high oxidation potential, which completely disrupts the synthesis of protein. Myhrstad and Sandal(12) have proposed a mechanism for the oxidation of biological material in water in the pH range of 6-8. In the first step chlorine dioxide is reduced to chlorite:

$$ClO_2 + e^- = ClO_2^- \qquad (5.12)$$

Chlorous acid is unstable except in very dilute solution, but to a limited extent it is associated ($k_a = 1.1 \times 10^{-2}$).

$$ClO_2^- + H^+ = HClO_2 \qquad (5.13)$$

Finally, chlorous acid disproportionates to chlorine dioxide and chloric acid:

$$4HClO_2 = 2ClO_2 + H^+ + Cl^- + HClO_3 + H_2O \qquad (5.14)$$

Chlorine dioxide reacts with hydroxyl ion to produce chlorite and chlorate ions:

$$2OH^- + 2ClO_2 = ClO_2^- + ClO_3^- + H_2O \qquad (5.15)$$

According to Latimer(7) this reaction goes readily in alkaline solution, and its free energy ($\Delta F = -38.3$ kcal) indicates that it should be reversed to give appreciable concentrations of chlorine dioxide in acidic solution, but experimentally chlorine dioxide is not formed under these conditions. Thus, there is some question as to just what the biologically active species is in aqueous solution.

1.2 Enzyme Poisons

Enzymes are large molecules that catalyze the numerous biological reactions that comprise the metabolism of an organism. Thus, cytochromes are essential to respiration, permeases regulate the permeability of the cytoplasmic membrane, and lysozymes catalyze the decomposition of moribund organisms. In addition, there are specific enzymes, such as the cellulases, which are secreted by cellulolytic microorganisms enabling them to metabolize cellulose as a source of carbon, and urease, which enables certain organisms to hydrolyze urea to carbon dioxide and ammonia. A characteristic of enzymes is their ability to bind themselves to the substrate upon which they are acting by forming some sort of unstable chemical complex, or physical bond. Some enzymes are entirely protein, but others are conjugates that contain a nonprotein component, or prosthetic group, in addition to the protein or apoenzyme.[13]

Cysteine, an amino acid containing a terminal thiol group, is essential to the biological activity of many enzymes.

$$\begin{array}{c} \quad\quad\quad NH_2 \\ \quad\quad\quad | \\ HS{-}CH_2{-}CH{-}COOH \end{array}$$

The thiol group is susceptible to oxidation, to precipitation, and to complex formation. Therefore, any chemical agent that participates in one of these reactions with thiol can disrupt the metabolism of living cells. Other chemicals can block electron transfer through iron atoms in cytochromes, and thereby paralyze respiration. In this section, several toxicants that kill microorganisms by attacking one or more of the over 2500 enzymes that have thus far been identified are discussed.

Methylene Bisthiocyanate

Methylene bisthiocyanate, $CNS{-}CH_2{-}CNS$, the methylene ester of thiocyanic acid, is a powerful toxicant to bacteria, algae, and fungi. In normal respiration trivalent iron accepts electrons from primary cytochrome dehydrogenase, but in the presence of methylene bisthiocyanate the thiocyanate fragment of the molecule blocks the transfer of electrons by reacting with ferric iron to form the weak salt $Fe(CNS)_3$, or in excess, the red complex $Fe(CNS)_6^{---}$. The dissociation constants of these two reaction products are given by Latimer[7]

$$Fe(CNS)_3 = Fe^{+++} + 3CNS^- \qquad K = 2.6 \times 10^{-7}$$

$$Fe(CNS)_6^{---} = Fe^{+++} + 6CNS^- \qquad K = 8.0 \times 10^{-10}$$

Ferric iron is thus inactivated, causing the immediate death of the cell. Because of this fundamental mode of attack, microorganisms containing cytochromes are unable to adapt to methylene bisthiocyanate under any circumstances. The chemical also kills anaerobic bacteria such as *Desulfovibrio*, for this organism too has an iron-containing cytochrome that is vulnerable to the thiocyanate fragment.

Methylene bisthiocyanate hydrolyzes rapidly above pH 8, and also is inactivated by dissolved ferric ion circulating in a cooling system, with which it forms the familiar red color of $(FeSCN)^{++}$. Thiocyanate ion is formed in strong alkali along with lesser amounts of cyanide and sulfide, but hydrogen cyanide is not formed as a decomposition product either in normal use or in storage. Methylene bisthiocyanate is not very soluble in water, and if applied dissolved in an immiscible solvent, such as petroleum naphtha, the partition coefficient is such that most of it remains in the solvent. If a water-soluble solvent is used the compound precipitates immediately upon adding the solution to water. Buckman and Pera[(14)] have disclosed the use of specific dispersants and penetrants in formulated liquid products containing methylene bisthiocyanate. A dispersant is used to achieve an effective concentration of the active ingredient throughout the water system; penetrants are included to invade the slime layers of bacteria and algae. Nonionic dispersants of the type $R\text{-}O\text{-}(R'\text{-}C_2H_3\text{-}O)_xH$ are suitable. These are polyalkyleneglycol ethers, or alkylphenylpolyalkyleneglycol ethers, depending on R and R′. In the simple case when $x = 2$, $R = C_2H_5$, and $R' = H$, the compound is $HO\text{-}(CH_2)_2\text{-}O\text{-}(CH_2)_2\text{-}O\text{-}C_2H_5$, dimethyleneglycolmonoethyl ether.

Dimethylamides with the formula $R\text{-}(CO)\text{-}N(CH_3)_2$ also are used in the proposed formulation. R may be hydrogen, or saturated, unsaturated, or hydroxy-substituted alkyl radicals containing from 1–22 carbon atoms; in the simplest case when R = H, the compound is dimethylformamide.

It would be impracticable to attempt to use methylene bisthiocyanate by itself; proprietary formulations such as that just described are much more satisfactory and economical. Low concentrations of the toxicant are effective (3–4 ppm), and because the molecule is nonionic, it is not inactivated by dirt, debris, or other particulate matter in the water system, and it easily penetrates the cytoplasmic membrane. Happold, et al.[(15)] have described an organism, which they assign to the *Thiobacillus* genus, that metabolizes thiocyanate to ammonia and sulfate; this organism, however, is not encountered in cooling water.[(16)]

Chloroethylene Bisthiocyanate

Siegel and Rosenblatt[(17)] have studied the release of thiocyanate ion from ethylene bisthiocyanate, and propose the following mechanism:

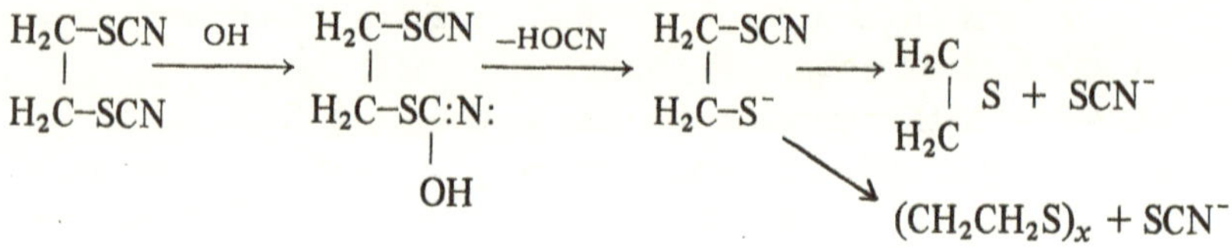

Here again, the thiocyanate ion combines with iron atoms in cytochromes situated in the cytoplasmic membrane, blocking the transfer of electrons that is essential to respiration. As with methylene bisthiocyanate, penetrants, dispersants, and solvents are used in commercial formulations of chloroethylene bisthiocyanate to increase the vulnerability of microorganisms to the active ingredient.

Acrolein

Acrolein, CH_2=CH-CHO, belongs to the class of organic compounds called alphamethylidine alkanals, or more commonly, alpha, beta-unsaturated aldehydes. It is a flammable, volatile, lachrymatory liquid with a sharp, irritating odor. In common with other aldehydes it oxidizes readily in air, forming peroxides and free-radical intermediates that initiate rapid polymerization to polyacrolein. If a small concentration of the reducing agent hydroquinone is added to monomeric acrolein it reacts with these free-radicals to inhibit polymerization. Repeating units in polyacrolein include the following:

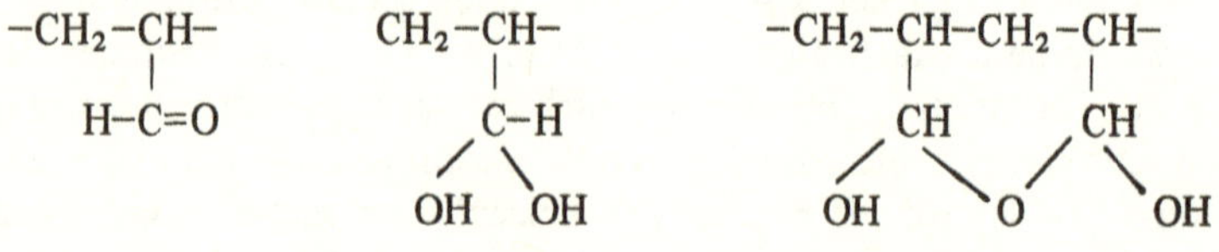

The compound possesses two sites capable of polymerizing: the carbonyl group and the double bond. Polyacrolein is insoluble in water and is inactive as a biocide. Stack[18] in measuring the toxicity of acrolein and related alpha,beta-unsaturated carbonyl compounds, found that the threshold toxicity of acrolein is 1.5 ppm to some unspecified microorganisms isolated from domestic sewage. He also found that microorganisms acclimate to acrolein, and that after one hundred days the threshold toxicity had risen to 18 ppm, an impracticably high concentration to maintain in cooling water systems.

Legator[19] has disclosed a procedure for controlling anaerobic bacteria in oil-field drilling operations using acrolein in the concentration range of 10–50 ppm. Special feeding equipment is required to apply acrolein safely, but once in the water it is innocuous to personnel. Because of its intense reactivity the compound does not persist. When acrolein is added to water hydration occurs immediately on account of the affinity of the conjugated system for electrons.

$$CH_2{=}CH\text{-}COH + H_2O = CH_2(OH)\text{-}CH_2COH \qquad (5.16)$$

This reduces its toxicity to some extent, and the analogous reaction with hydrogen sulfide deactivates the compound completely.

$$CH_2{=}CH\text{-}COH + H_2S = CH_2(SH)\text{-}CH_2COH \qquad (5.17)$$

Acrolein is toxic to microorganisms because a similar reaction with thiol groups disrupts metabolism. The compound forms two adducts of bisulfite.

$$CH_2{=}CH\text{-}CHO + NaHSO_3 = CH_2{=}CH\text{-}CH(OH)SO_3Na \qquad (5.18)$$

$$CH_2{=}CH\text{-}CH(OH)SO_3Na + NaSO_3 = NaSO_3CH_2\text{-}CH_2\text{-}CH(OH)SO_3Na \qquad (5.19)$$

The first product is sodium 1-hydroxy-2-propene-1-sulfonate; the second is sodium 1-hydroxy-1,3-propanedisulfonate.

Heavy Metals

Salts of heavy metals form insoluble mercaptides and precipitate protein; they are intensely toxic to microorganisms and all other forms of life but, in general, are considered to be bacteriostats. In sanitary engineering copper sulfate at a concentration of about 1 ppm has been used for many years to eradicate algae from reservoirs and ponds. Susceptibility varies a great deal with different algal species, but molds, fungi, and bacteria are completely unaffected by low concentrations of copper ion. Copper citrate is used sometimes as a microbicide in cooling systems, although it is eminently inappropriate for this purpose. Cupric ions plate on steel, become a cathode and the surrounding steel corrodes. In this coupled system iron gives cathodic protection to the copper and is itself sacrificed. Furthermore, citrate is known to stimulate algal growth and is the preferred source of carbon for certain *Aerobacter.*

It is presumed that the toxic effect of cupric ion is achieved by coagulating colloids in the bacterial cytoplasm. Stable chelates are formed, however, with many natural amino acids including alanine, leucine, valine, and glycine. The dissociation constant of the latter is given by Latimer(7) as $K = 7.9 \times 10^{-16}$. The structure of the copper chelate of glycine is

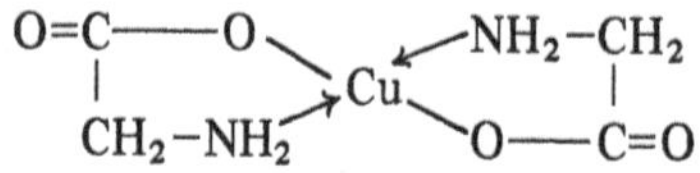

In general, organic compounds of heavy metals are un-ionized, and thus pass through the cytoplasmic membrane with greater ease than do the inorganic salts. The organomercurials phenylmercuric acetate and methylmercury are for this reason more toxic to living cells than are inorganic mercury salts; by attacking thiol groups they cause extensive irreversible derangement of metabolism.

With the exception of an organic tin compound, heavy metals are seldom used for the microbiological control of cooling water, and their use in other applications is being curtailed rapidly on account of their toxicity to fish and other animal life. Tin is most often applied to cooling water as the distannoxane, bis(tri-*n*-butyltin)oxide,

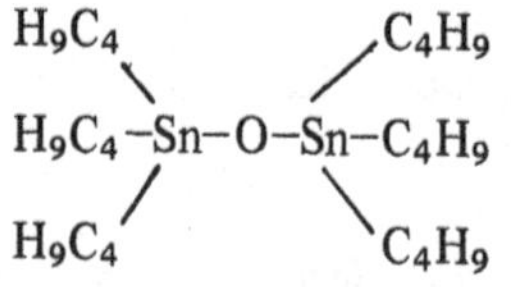

in formulations stabilized with a quaternary ammonium salt to inhibit the precipitation of meta-stannic acid, and so extend shelf-life. For a typical formulation see Table 4.4. Tin functions as a biocide by interfering with oxidative phosphorylation, but for some reason it is more effective against Gram-positive than Gram-negative bacteria. As this compound is extremely toxic care must be taken in disposing of blowdown water from cooling systems wherein it is used to avoid poisoning the microorganisms in secondary waste treatment plants, or fish and other aquatic life.

1.3 Toxicants that Disrupt the Cell Wall and Cytoplasm

A group of chemicals that achieve a toxic effect by virtue of their surface activity is now considered. Surface active agents are organic chemical salts containing a hydrophilic and a balancing hydrophobic portion in the same molecule. Thus, cetylpyridinium chloride, $C_{16}H_{33}(C_5H_5N)^+Cl^-$, is cationic (positive), and dodecyl sodium sulfate, $CH_3(CH_2)_{11}OSO_3^-Na^+$, is anionic (negative). As the anionic compounds are effective only against Gram-positive organisms, they are of no use in cooling water. The toxic effect of all chemicals discussed in this section depends upon their ability to orient at the surface of the cytoplasmic membrane where they produce cytolytic damage. In contrast to the enzyme poisons, which are toxic at low concentration, relatively high concentrations are required of chemicals that attack the cell walls, as their action to a large degree is mechanical.

Dodecylguanidine Hydrochloride

The microbicidal properties of long-chain aliphatic guanidines and their salts have been described by Pelcak and Dornbush.[20] Dodecylguanidine hydrochloride is particularly valuable for treating cooling water, for it not only kills algae and bacteria, it also forms an electrostatic bond with carboxyl groups in cellulose, thus protecting wooden cooling towers from invasion by fungi. Also, the scouring effect of this powerful surfactant maintains exceptionally clean cooling towers. Insoluble salts are formed with inorganic anions such as phosphate, sulfate, and chromate when these ions are present at sufficient concentration. The chromate salt, however, does not form at concentrations normally used in cooling water, i.e., 10–25 ppm CrO_4.

Dodecylguanidine, in common with other cationic surfactants, ruptures the loose hydrogen bond structure that joins protein to coenzyme in many enzyme systems. The formula of the hydrochloride is

$$n\text{-}C_{12}H_{25}NH\text{-}\overset{\overset{\displaystyle NH}{\|}}{C}\text{-}NH_3^+Cl^-$$

Acid salts are invariably used for they are water-soluble. Additional advantages of this chemical are its biodegradability, its dispersing action on many inorganic salts and oxides, and its ability to bring into solution bacteria clinging to surfaces. As with all surfactants, dodecylguanidine alters the surface tension on the gill surfaces of fish, interfering with the transport of oxygen and destroying the animal.

Chlorophenols

The ability of phenols and chlorophenols to orient at interfaces plays a significant role in their toxicity to microorganisms. Phenols combine with the cell wall and cytoplasmic membrane by an adsorption process, probably hydrogen bonding, rather than by chemical reaction. After adsorption and infiltration of the cytoplasmic membrane, phenols do lethal damage by forming a colloidal solution with the cytoplasm, and precipitating protein. In contrast to chlorine and most surface-active chemicals the presence of excessive amounts of organic material has little effect on the efficiency of phenols. Spores resist phenol, but sporulating bacteria (*Bacillus*, *Clostridium*) are seldom found in troublesome numbers in cooling water. Pseudomonads, however, also develop resistance, and in addition, many of these species adapt quickly, producing enzymes that enable them to metabolize normally toxic chemicals including cresols, phenols, chlorophenols, and nitrophenols. Growth is often inhibited, but the organisms continue to respire, after which they may begin to grow prolifically as they adapt to the toxicant. Two of the fungi that cause soft-rot, *Chaetomium globosum* and *Polyporus versicolor*, are especially tolerant to phenols and chlorophenols.

Sanborn[(21)] has published the results of a study of the effect of various phenolic compounds and chlorinated phenols on certain genera of *Aerobacter* and other coliform slime-producers; unfortunately the toxicants are not specifically identified. Mixtures of tri- and pentachlorophenols are up to twice as active as either chemical alone, and the biological activity can be further increased by adding a small amount of an anionic surfactant such as dodecyl sodium sulfate, which by lowering interfacial tension, increases the rate of penetration of the cytoplasmic membrane. Pentachlorophenol, which penetrates wood better than the corresponding sodium salt (pentachlorophenate), is used to protect cooling tower lumber against the invasion of cellulolytic fungi.

Chlorinated phenols, being highly toxic to fish, are especially objectionable in effluents discharged into lakes, rivers, or oceans. They impart offensive taste and odor to drinking water at very low concentration (3–5 ppb), and being lipophilic, dissolve in the fatty tissues of fish that ingest them, strongly flavoring the flesh of the animals, even at sublethal dosages. A further objection to chlorophenols is the possible presence of a contaminant, tetrachlorodibenzo-para-dioxin, which is believed to have caused birth deformities in mice exposed to 2,4,5-trichlorophenol. Because of these serious objections, the manufacture of chlorophenols is being discontinued in the United States, and before long none of these compounds will be available.

Quaternary Ammonium Salts

Quaternary compounds owe their biological activity to surface-active properties, or electrostatic attraction to negatively-charged receptor groups in the cell wall and cytoplasmic membrane. The general subject of surface active agents and their effect on microbial cells has been thoroughly covered in a monograph published by the Society of Chemical Industry of London.[22] In this monograph James[23] has outlined a theory of antibacterial activity of quaternary compounds and other surfactants. Quaternary ammonium salts are organically substituted nitrogen compounds in which the nitrogen atom is pentavalent. Four of the substituents are alkyl, aryl, or heterocyclic radicals; the fifth is an anion. These salts are analogous to ammonium chloride, in which organic radicals substitute for one or more of the hydrogen atoms. To be active as a biocide at least one of the organic radicals must have eight to eighteen carbon atoms in the chain.[24] A typical quaternary compound is *N*-dodecylbenzyl-*N*,*N*,*N*-triethylammonium chloride.

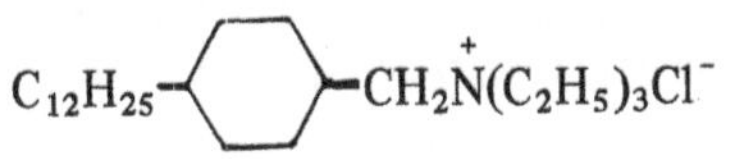

Sexton[25] states that cationic exchange by bacteria is the basis for the activity of metallic salts, basic dyes, and cationic surface-active agents. McCalla[26] also has studied cationic adsorption by bacteria, all of which carry a negative charge in neutral or alkaline solution. For this reason adaptation is an especially common phenomenon in cooling water maintained at low pH, i.e., 6.0–6.5. Quaternary compounds are practically without effect on spores, but they possess some nonspecific fungicidal activity in the relatively high concentration range of 100–300 ppm. In use they are prone to foam, but this is usually harmless.

1.4 Miscellaneous Microbicides

Numerous mixtures of chemicals have been given proprietary names, and offered for sale as microbicides, although few of these have any advantage over the chemicals already discussed for treating cooling water. There are several additional compounds, however, that have merit as toxicants for algae, bacteria, and fungi. Derivatives of dithiocarbamic acid, NH_2CSSH, such as zinc dimethyldithiocarbamate, $Zn[(CH_3)_2N\text{–}CSS]_2$, are especially active, but have the disadvantage of being persistent and extremely toxic to fish. Dithiocarbamates are chelating agents that combine with metal ions in enzymes essential to the metabolism of microorganisms. The sodium salts are also used. Certain alkylamines are highly toxic, particularly when combined with chlorophenols. Two effective compounds of this type are *N*-alkyl-1,3-propylenediamine, and the acetate of 1-alkylamino-3-aminopropane; the alkyl groups are in the C_6-C_{18} range. Cooling systems are occasionally infested with *Aspergillus* molds that are difficult to eradicate. A recently introduced chemical, 2,2-dibromo-3-nitrilopropionamide, is unusually effective against these molds (and also against *Desulfovibrio desulfuricans*) in the rather low concentration range of 10-15 ppm. The proper name for this chemical is 2,2-dibromo-3-cyanoacetamide.

The sodium salt of 2-mercaptobenzothiazole is a fairly efficient corrosion inhibitor that shows some toxicity to microorganisms. At the small concentrations (1-2 ppm) used to inhibit corrosion, however, pseudomonads quickly adapt to the chemical as a carbon source. When used as a biocide sodium 2-mercaptobenzothiazole is always combined with some other agents, but as it begins to precipitate at pH 6.6 it is not really useful as a biocide below pH 7.

Although, in general, toxic compounds do not differ markedly in their action on algae, bacteria, and fungi, Gratteau[27] has described an intriguing new algicide, 2,3-dichloronaphthoquinone, which is toxic to blue-green algae at 25-50 ppb, but not to green algae, fish, aquatic plants, or zooplankton. The reason for this remarkable selectivity is not known.

2. CHEMICAL STRUCTURE AND TOXICITY OF MICROBICIDES

The possibility of developing selectively toxic chemicals using the principles of comparative biochemistry appeals strongly to the logical mind, yet it is quite common for compounds having similar chemical structures to vary greatly in toxicity. Methylene bisthiocyanate, for example, is a highly toxic enzyme poison, whereas methylthiocyanate is virtually worthless as a microbicide, being about one two-hundredths as toxic as the bisthiocyanate. Similarly, Stack[18] found that the threshold toxicity of acrolein, CH_2=CHCHO, is 1.5 ppm, while that of propionaldehyde, CH_3CH_2CHO, is greater than 500 ppm.

Despite these apparent anomalies there are many well-established correlations between chemical structure and biological activity. Sexton[25] has discussed these in detail, and Hugo[28] has published an informative account of the mechanism of action of numerous antibacterial agents. Kiser, et al.[3] have cited six lethal effects that control the growth of microorganisms

1. Coagulation of protein.
2. Inactivation of enzymes.
3. Disruption of cellular lipids.
4. Damage to genetic apparatus.
5. Destruction of ribonucleic acid.
6. Damage to cell walls.

Protein is coagulated by phenols, chlorophenols, copper, zinc, and other heavy metal ions. Methylene bisthiocyanate and chloroethylene bisthiocyanate inactivate iron in cytochrome respiratory enzymes; acrolein inactivates cysteine; mercury precipitates enzymes that contain thiol groups; and chlorine oxidizes thiols. Chlorine dioxide specifically halts protein synthesis,[29] perhaps by attacking ribosomes in the cytoplasm, or else by inactivating ribonucleic acid. Both dodecylguanidine hydrochloride and bis(tri-*n*-butyltin)oxide interfere with oxidative phosphorylation, a vital process in cellular respiration. Lipid components of cells are especially vulnerable to heat, and also to various cationic surfactants that degrade lipids. Deoxyribonucleic acid regulates the production of enzymes in all living cells, so chemicals that damage this regulator (chlorine, cationic surfactants that break hydrogen bonds, and various antibiotics) are lethal to cells. The so-called genetic messenger, ribonucleic acid, is similarly damaged by chlorine, cationic surfactants, and the antibiotic streptomycin. Deoxyribonucleic acid is found in the bacterial chromosome, while ribonucleic acid is mostly in the cytoplasm with smaller amounts in the chromosome. Intracellular penetration by phenols, chlorophenols, quaternary ammonium salts, and dodecylguanidine hydrochloride lyse the mucopeptide layer in the cell wall, causing extracellular leakage of the contents of the cell and death of the organism. We now give a few more particulars of the chemistry of certain of the microbicides considered in the previous section.

2.1 Oxidizing Agents

As mentioned before, thiol groups, as components of apoenzymes, are vital to the functioning of many metabolic processes. Chlorine oxidizes the thiol cysteine to the disulfide cystine, thus inactivating numerous enzymes.

$$2SHCH_2CHNH_2COOH + Cl_2 = \begin{array}{l} SCH_2CHNH_2COOH \\ | \\ SCH_2CHNH_2COOH \end{array} + 2H^+ + 2Cl^- \qquad (5.20)$$

Furthermore, the amino groups are chlorinated to chloramines, shifting the isoelectric point of the amino acid, and generally devastating the cytoplasm.

$$\begin{array}{l} SCH_2CHNH_2COOH \\ | \\ SCH_2CHNH_2COOH \end{array} + 4Cl_2 = \begin{array}{l} SCH_2CHNCl_2COOH \\ | \\ SCH_2CHNCl_2COOH \end{array} + 4H^+ + 4Cl^- \qquad (5.21)$$

Chlorine dioxide does not halogenate the amino groups, but is capable of oxidizing the disulfide to the corresponding sulfoxide, and ultimately, to the sulfone.

TABLE 5.2

Effect of Temperature on Disinfection by Chlorine Dioxide

Chlorine Dioxide (ppm)	*Temperature* (C)	*Time for 99 percent Kill* (s)
0.25	5	110
0.25	10	74
0.25	20	41
0.25	30	16
0.75	5	60
0.75	30	14

Benarde, et al.,[29] in studying the kinetics and mechanism of disinfection by chlorine dioxide, found that its killing efficiency varies directly with temperature, suggesting that the rate of kill is controlled by a chemical reaction rather than by some physical effect. A well-known principle of chemistry is that an increase in temperature of 10 C generally doubles or triples the rate of a chemical reaction.

The extremely rapid rate of kill suggests that chlorine dioxide breaches the cell wall, but this is not the case. The authors verified that the chemical does not lyse cells, nor does it oxidize protein, amino acids, or nucleic acids. As already mentioned the chemical exerts its lethal effect on microorganisms by interrupting the synthesis of protein.

2.2 Enzyme Poisons

The toxicity of the alkylthiocyanates is considerably influenced by their physical properties, notably solubility, and also by the alkyl group. Thus, Wehner and Hinz[17] report that methylene bisthiocyanate, $SCN\text{-}CH_2\text{-}SCN$, and vinylene bisthiocyanate, $SCNCH{=}CHSCN$, are both highly toxic to *Bacillus mycoides* and *Aspergillus flavus*, whereas ethylene bisthiocyanate, $SCNCH_2CH_2SCN$, and chloroethylthiocyanate, CH_3ClCH_2SCN, are less than one-twentieth as active. On the other hand, chloroethylene bisthiocyanate, $SCNCHClCH_2SCN$, is two and one-half times as toxic as methylene bisthiocyanate to the same two organisms. Sexton[25] points out that the methyl group is unique among alkyl groups because of the special role it plays in metabolism. Either transmethylation or the replacement of active hydrogen atoms by methyl groups releases thiocyanate fragments within the cell, which then paralyze the respiratory process by combining with iron in cytochromes, thereby blocking the transfer of electrons. Despite this, methylthiocyanate, CH_3CNS, is worthless as a microbicide.

Thiols have a strong affinity for double bonds, particularly those chemically activated by a neighboring ketone group, as in acrolein. Thus, cysteine reacts readily with acrolein to form a compound similar to that formed with hydrogen sulfide.

$$CH_2{=}CHCO + SHCH_2CHNH_2COOH = CH_2(SCH_2CHNH_2COOH)CH_2CHO \qquad (5.22)$$

Other toxic α,β-unsaturated compounds are meth-acrolein, $CH_2{=}C(CH_3)CHO$, and methylvinyl ketone, $CH_2{=}CHCOCH_3$. Although neither of these is sold as a microbicide for application to cooling water, both have the distinct advantage that bacteria cannot acclimate to them.[18]

The heavy metals, of which mercury serves as an example, form insoluble mercaptides, thereby inactivating enzymes that contain thiol groups.

$$2SHCH_2CHNH_2COOH + Hg^{++} = Hg\begin{cases} SCH_2CHNH_2COOH \\ SCH_2CHNH_2COOH \end{cases} + 2H^+ \qquad (5.23)$$

2.3 Toxicants That Attack the Cell Wall

Cationic surface active compounds, such as the acid salts of dodecylguanidine and the quaternary ammonium salts, form electrostatic bonds with carboxyl groups in proteins and enzymes that interfere with oxidation–reduction and other biochemical reactions. The basic (cationic) surfactants are better able to penetrate cell walls than are the acidic (anionic) ones, and thus more efficient, particularly against Gram-negative bacteria. In this way the cell wall is damaged, lysis occurs, and metabolites leak out of the cell.

Substituted guanidines are prepared by reacting salts of primary alkylamines with cyanamide in isopropanol.[20] Maximum toxicity is attained with 10–14 carbon atoms in the alkyl group; toxicity is a function of chain length rather than of molecular weight, and is decreased by unsaturation in the chain. The respiratory enzymes that catalyze oxidative phosphorylation, the source of most of the energy that sustains metabolic processes in living cells, are associated with membranes called mitochondria in plant and animal cells. Although bacteria do not have mitochondrial membranes they possess similar enzymes, differently arranged, that are able to carry out the same chemical reactions. Oxidative phosphorylation is discussed further on page 117. Pressman[30] has shown that guanidine and its derivatives, including dodecylguanidine, are blocking agents of energy transfer between the electron transport chain leading from substrate to oxygen and adenosine triphosphate. In this manner dodecylguanidine strongly inhibits bacterial respiration in addition to its surface effect on cell walls. Thus, it could also be classified as an enzyme poison. Under the trade name Dodine the chemical has been used for many years as a plant fungicide, functioning as such by preventing the germination of spores.

Albert[31] reports that electron photomicrographs reveal that quaternary ammonium salts rapidly dissolve large areas of the cell wall, liberating almost the entire contents of the cell within a few minutes. Surface active agents concentrate on the surface of cells where they break ionic bonds, thus shattering the structure of the cell. In the homologous series of quaternary ammonium salts, $R(CH_3)_3N^+$, where R is C_5H_{11}, or higher, the salt exhibits antiseptic properties, which reach a maximum between C_{12} and C_{14}. These salts dissociate conjugated proteins and lipo-proteins in the cytoplasmic membrane, disturbing this osmotic barrier. If the cell wall is also damaged, lysis occurs by osmotic explosion, i.e., the cell swells until the weakened wall ruptures, and the organism is destroyed. The technical term for this phenomenon is plasmoptysis.

Chelating agents such as the dithiocarbamates form undissociated complexes with certain metal ions essential to biological activity. Examples of these are magnesium in chlorophyll, iron in cytochromes, copper in porphyrins and albumin, calcium, magnesium, and zinc in hydrolytic enzymes, and cobalt in vitamin B_{12}. Most of these metals are made unavailable for biochemical reactions by chelating agents, which for this reason are toxic to microbes.

3. *EVALUATION OF MICROBICIDES*

Toxicant evaluations are made by manufacturers and vendors to screen and rank chemicals used, or proposed, as biocides. This is done by subjecting a portion of cooling water containing 5.0×10^5-3.0×10^6 bacteria per milliliter to known concentrations of microbicide for a certain number of hours. At the end of the specified period dilutions are made, and the number of bacteria surviving is determined by making a viable plate count according to the procedure in Chapter IX. Toxicant evaluations are customarily expressed as "percent kill," which is misleading; the biological cleanliness of a cooling system is not a function of "percent kill," but of residual bacteria, that is, "percent survivors." Compare, for example, the results tabulated in Table 6.12, with those in Table 9.2. The cost of maintaining microbiological control of a cooling system can be a substantial percentage of the total cost of chemically treating the water, but when comparing the relative costs of using different biocides the unit cost (price per pound) is only one factor. It is also necessary to determine the minimum effective dosage of each toxicant being evaluated, and also the duration of its effect. In Chapter VI, Section 5.3, we show how to use toxicant evaluation as a quantitative guide for choosing the minimum effective dosages of certain microbicides.

If facilities are unavailable for making toxicant evaluations, a useful aid for judging microbicides is a review of relative toxicities of biostatic agents, which until 1969 was published every five years by the Technical Association of The Pulp and Paper Industry.[32] In these reviews proprietary biocides are listed by trade name, their active ingredient is identified, and the concentration of each required to inhibit the growth of four different microorganisms are tabulated. The publication of this revealing information has now been discontinued.

4. *FACTORS AFFECTING THE EFFICIENCY OF MICROBICIDES*

The performance of a toxic chemical in an evaluation in the laboratory is seldom equaled in an operating cooling system because contaminants and extraneous matter in the water act in one way or another to protect microorganisms, or otherwise interfere with microbicides. Thus, the presence of sources of organic carbon protect algae, and raise the minimum lethal dose above that measured in an inorganic medium. Other influences on the efficiency of a toxicant include the pH of the recirculating water, the concentration of soluble inorganic salts, dirty screens, chemical contaminants, oil, biological debris, and other miscellaneous insoluble material in the cooling system.

Chlorine has little penetrating power, and thus is ineffective when biological debris such as algal mats and slime are present. It is also inactivated by easily chlorinated contaminants including ammonia, ethanolamines, and phenols. Furthermore, chlorine is essentially nontoxic at pH 8, or higher. Chlorine dioxide, on the other hand, is much more persistent, and more lethal than chlorine to both spores and vegetative cells, without suffering any of the limitations of chlorine.

Methylene bisthiocyanate is inactivated in systems contaminated by oil leaks, as the partition coefficient favors its passage into the hydrocarbon phase, where it is innocuous to microorganisms. The chemical hydrolyzes rapidly above pH 8, and is uneconomical to use above pH 7.5. It is, therefore, unsuitable for nonchromate treatments, which almost without exception are alkaline; chloroethylene bisthiocyanate is appropriate for these treatments.

High concentrations of cationic surfactants, such as dodecylguanidine hydrochloride and quaternary ammonium salts, induce the agglomeration of bacteria into masses, which then attach to surfaces for physical protection against exposure to a lethal dose. Quaternary compounds and other cationic agents are inhibited by anions, which neutralize positive electrical charges. Therefore, they are easily inactivated by the profusion of negatively charged materials in cooling systems, which includes inorganic salts, oxides, wood, and other organic matter. Dirty screens are highly efficient adsorbers of cationic microbicides, reducing their effective concentration in the water to such a level that they actually stimulate the growth of pseudomonads, which quickly adapt to these agents as a source of carbon. A relatively short time after being added to cooling water the concentration of cationic surfactants falls to 10 percent or less of the initial slug dose. As most of the chemical remains in the cooling system attached to negative sites in the tower or on microbiological films, very little escapes in the blowdown water. For this reason, it is pointless to apply cationic surfactants to cooling systems with voluminous fouling, dirty screens, or a tower basin containing heavy sludge. Phospholipids, proteins, and nucleic acids all form complexes with surfactants, which thus modify or destroy the structure of both the cell wall and the cytoplasmic membrane.[33] The maximum kill for a particular concentration of surfactant probably occurs within the first few minutes after adding the biocide. There is also evidence that high concentrations of calcium and magnesium reduce the toxicity of quaternary compounds by competing with them for acidic sites in the cell walls of bacteria. As mentioned, many bacteria acclimate rather quickly to these compounds, especially at pH values less than seven.[35]

Chlorophenols are insensitive to pH in the range of 5.5-8.5, and do not suffer any of the other interferences mentioned, although other considerations cited above in Section 1 often preclude their use.

5. *SPECIAL CONSIDERATIONS*

Three topics are discussed in this section that do not conveniently fit in any other part of the book. First, the very important subject of safe handling of microbicides is addressed. The second topic, dealing with methods for preserving wooden cooling towers, is not directly related to cooling water treatment, but nevertheless, is an important subject that demands some attention. Finally, the compatibilities of microbicides with various other chemicals used to treat cooling water are briefly discussed.

5.1 Safety Precautions in Handling Microbicides

All commercial formulations of biocides are hazardous chemicals that must be handled and dispensed with great care to avoid injury. A treadle-operated safety shower and an eye-wash fountain should be installed in areas where these chemicals are dispensed. Regardless of the chemical being handled, the eyes must be protected with approved chemical goggles; safety spectacles are not sufficient. Wearing clean clothing that covers the arms and legs completely, protecting the hands with rubber gloves, and making certain that there is adequate ventilation in the work area are essential. If clothing becomes contaminated by splashes or spills of a biocide, the clothing must be removed immediately, then they must be washed thoroughly with soap and water before being worn again. When barrels containing biocides are emptied, they should be washed out thoroughly before discarding. As a rule, these barrels should not be reused. Spills of liquid biocides can be cleaned up by absorbing them on attapulgus clay, Fuller's earth, or diatomaceous earth, after which the material must be disposed of in an approved dumping area. Other personal safety equipment should include rubber gloves, rubber boots, aprons and face shields; if chlorine or acrolein is used, a chemical cartridge respirator or a self-contained breathing apparatus should be available for respiratory protection in the event of leaks or other failure of equipment.

In Table 5.3 some of the harmful properties of a number of commonly used microbicides are listed, together with their usual concentrations in commercial formulations. The values tabulated under the heading Oral LD_{50}, mg/kg are median lethal dosages—those that are fatal to 50 percent of the test animals, in this case rats. Mice, guinea pigs, and rabbits are also used for determining this value. Most of the toxicants listed in Table 5.3 are classified as primary irritants and sensitizers. In addition, bis(tri-*n*-butyltin)oxide, methylene bisthiocyanate, and sodium dimethyldithiocarbamate depress the central nervous system. Symptoms of poisoning include drowsiness, fatigue, headache, unconciousness, narcosis, and convulsions. Pentachlorophenol, if inhaled, ingested, or absorbed through the skin, increases the basal metabolism rate, and interferes with oxidative phosphorylation.

Plunkett[(34)] describes the pathology, symptoms, diagnostic tests, and treatments used in cases of poisoning by a large number of industrial chemicals. Prompt first aid is essential, particularly when the eyes are affected. They should be irrigated with clear water for at least 15 min, rolling back the eyelid to facilitate the removal of chemical, after which the accident victim should be examined by a physician. Spills on the skin should be washed off immediately and thoroughly with soap and water. If the skin

TABLE 5.3

Toxicity of Selected Microbicides

Microbicide	*Usual Concentration* (%)	*Absorbed Through Skin*	*Causes Conjunctivitis*	*Causes Corneal Damage*	*Skin Burns*	*Oral* LD_{50} mg/kg
Acrolein	90	+		+	+	40
Bis(tri-*n*-butyltin)oxide	5	+	+	+	+	190
Chlorine	100			+		—
Chlorine Dioxide (stabilized)	6		+	+	+	—
2,2-Dibromo-3-nitrilopropionamide	20		+		+	130
Dodecylguanidine hydrochloride	35	+	+			700
Methylene bisthiocyanate	10	+	+	+	+	230
Pentachlorophenol	100	+	+	+	+	200
Quaternary Ammonium Salts	50	+	+			2000
Sodium Dimethyldithiocarbamate	30	+	+			500

is reddened, blistered, or ulcerated, consult a physician. If any of the chemicals in Table 5.3 are ingested, call a physician immediately. Should dodecylguanidine hydrochloride or methylene bisthiocyanate be accidentally swallowed, demulcents such as milk, olive oil, or egg whites can be given, if immediately available, otherwise give large volumes of water to dilute the toxicant. Do not induce vomiting, as it is likely to cause considerable mucosal damage. With the other toxicants the physician will usually induce vomiting then perform gastric lavage, followed by saline catharsis. It is unlikely that anyone would ingest any great quantity of a microbicide, so if the victim survives the acute stage, there is seldom any permanent disability following poisoning by any of the chemicals under discussion. In general, medical treatment is symptomatic and supportive, but nonetheless a physician should always be consulted in suspected cases of poisoning. Physicians can obtain advice on medical treatment for poisoning by specific toxic chemicals by calling Chemtrac in Washington, D.C., at 202-483-7616.

A number of toxicants owe their ability to penetrate the cytoplasmic membranes of microorganisms to their nonpolar or nonionic chemical structure. This also renders them insoluble, or very slightly soluble, in water, so they are formulated with dispersants, penetrants, wetting agents, and solvents to achieve uniform distribution in cooling water. These additives and solvents are in most instances themselves harmful to the human body. Isopropanol is used as a solvent in many commercial formulations of biocides. If splashed in the eyes it causes conjunctivitis and corneal necrosis, it produces dermatitis if spilled on the skin, and if ingested, nausea and vomiting ensue. The solvent is metabolized in the body to acetone, which is a depressant of the central nervous system, and as such, may cause dizziness, headache, and hypotension, among other symptoms. Dimethylformamide and ethyleneglycol monomethyl ether are used in some formulations of methylene bisthiocyanate to enhance its penetration. Both of these additives cause liver and kidney damage if ingested. Also, both are irritating to the eyes. Dimethylformamide is rapidly absorbed through the skin, leading to nausea, vomiting, and loss of apetite. Although not specifically skin irritants, anionic wetting agents (e.g., sodium dioctylsulfosuccinate) with repeated or prolonged exposures, remove natural oils resulting in dryness and fissuring of the skin; they are also severely irritating to the eyes.

Cooling towers are protected from fungal attack by spraying the wood with a solution of sodium pentachlorophenate. This inherently hazardous operation is carried out one cell at a time with careful attention to the protective equipment of the workmen and to the proper disposal of blowdown water containing pentachlorophenate. This chemical is highly persistent and intensely poisonous to bacteria and other microorganisms that flourish in secondary waste treatment processes. As spraying requires special equipment and elaborate safety precautions, it had best be left to an experienced contractor.

Under anaerobic conditions, as in the mud in the basin of a cooling tower, a substantial level of biocide may accumulate, shielded from normal biological degradation. For this reason, workmen engaged in cleaning basins should wear adequate protective equipment to prevent dermatitis or blistering of the skin. This very real hazard is frequently overlooked.

5.2 Preservation of Wood in Cooling Towers

In the course of time parts of the wooden structure of a cooling tower deteriorate from attack by fungi that metabolize cellulose, and flourish within the wood, where they cannot be reached by biocides in the recirculating water. Anyone who has attempted to cure an infection of ringworm or athlete's foot knows from the experience that fungi are extremely successful in coping with toxicants. Genera that invade the skin belong to the class *Hyphomycetes*, and include *Trichophytum*, *Microsporon*, *Endermophyton*, and *Epidermophyton*. Fungi that penetrate wood in cooling towers are similarly difficult to reach and kill, for as noted earlier, they have stout cell walls and high osmotic pressure, and although their cytoplasmic membrane is easily penetrated by lipophilic molecules (phenol, for instance) very few specifically antifungal agents are known. Indeed, once wood has been invaded it must be replaced, as it is impossible for chemicals to penetrate to the interior of structural members. Flooded sections such as the fill packing are subject to soft-rot, which is usually caused by *Fungi Imperfecti* or, less often, by *Ascomycetes*. Places that are occasionally, or partially, wet, for example, the heavier structural members, cell partitions, and drift eliminators, are invaded by *Basidiomycetes* that produce white-pocket-rot, whereas the relatively drier fan stacks, decking, and intake louvers may be attacked by still other *Basidiomycetes* that cause brown-rot.

Untreated clear heart redwood is naturally resistant to fungal infection on account of the presence of toxic chemicals in the wood; in most instances of premature decay, chemicals in the water have played an important role in accelerating the loss of these naturally protective extractives.

Decay can be significantly reduced by maintaining the pH of the recirculating water near seven, and by limiting chlorine residuals to about 0.8 ppm.

Baechler, et al.[36] have evaluated the performance of six wood preservatives by exposing slats impregnated with each for a period of five years in an operating cooling system. The preservatives were: (1) coal tar creosote, (2) acid copper chromate, (3) ammoniacal copper arsenite, (4) chromated copper arsenate, (5) chromated zinc arsenate, (6) pentachlorophenol (5 percent solution in heavy petroleum solvent). At the end of five years it was concluded that coal tar creosote was a more effective preservative than the heavy-metal formulations. The latter are applied as aqueous slurries. Acid copper chromate, for example, is prepared by mixing individual solutions of copper sulfate and sodium dichromate; the insoluble toxic salt, copper chromate, is deposited when the slurry is forced into lumber under pressure. In these same tests chromated copper arsenate proved least susceptible to leaching from the wood while in service.

Gjovik, et al.,[37] in a follow-up study of the same specimens after ten years in service, concluded that the water-based preservatives provided more protection than the two oil-based preservatives. Wood preservation is least successful for redwood and flat-grain Douglas fir, and most effective for ponderosa pine and western hemlock.

5.3 Reactions Between Biocides and Water-Treating Chemicals

At this point it is worthwhile mentioning a few of the common treatments used to inhibit scaling and corrosion in cooling systems, and to call attention to some undesirable reactions, or effects, of certain microbicides on these treatments.

In systems where chromate is used to inhibit corrosion the chromate ion itself exerts a significant leveling effect on the growth of microorganisms. For this reason otherwise inferior biocides may perform acceptably in chromate-treated water. When using this inhibitor, however, it is essential to keep algal growth under control, because algae secrete certain soluble extracellular substances into the water, e.g., glycolic acid and polysaccharides, that can reduce chromate, and also serve as sources of carbon for bacteria. Green algae and diatoms predominate in chromate-treated water.

Dithiocarbamates are incompatible with chromate as they rapidly reduce the ion and promote corrosion. Quaternary ammonium salts and dodecylguanidine acid salts form insoluble chromates if their local concentrations are high. Therefore, concentrated chromate solutions should not be added to a cooling system at the same time as either of these biocides, for they will mutually precipitate. At concentrations normally used, however, say 25 ppm of each chemical, chromate is not precipitated.

Careless application of chlorine produces excessively low pH, which rapidly destroys protective films of chromate. Sodium 2-mercaptobenzothiazole, a corrosion inhibitor in some nonchromate treatments, is oxidized by chlorine, chlorine dioxide, and 2,2-dibromo-3-nitrilopropionamide to the corresponding disulfide, which is insoluble in water and inactive as a corrosion inhibitor. This salt is also used in a few proprietary mixtures of microbicides, but as it precipitates below pH 6.6, it is inapplicable in a great number of water treatment programs. Similarly, methylene bisthiocyanate is susceptible to hydrolysis in slightly alkaline solutions, so cannot be used above pH 7.5.

Chelating agents, usually the sodium salts of nitrilotriacetic acid and ethylenediaminetetraacetic acid, find limited application in schemes for on-stream cleaning of cooling systems. As an incidental benefit, they exhibit some bactericidal effect by forming complexes with metal ions essential to metabolism, making them unavailable for biochemical reactions. Conversely, chelating agents also inactivate toxic cations, notably copper and zinc, enabling algae to grow in the presence of otherwise toxic concentrations, and these agents also prevent the precipitation of phosphate, both of which reactions promote the growth of microorganisms. Polyphosphates and phosphate esters, both of which hydrolyze to orthophosphate, also promote the growth of a great many microorganisms, making microbiological control more difficult than in systems from which these phosphorus compounds are absent. Phosphonates have no appreciable effect on microorganisms in cooling water as they are extremely stable toward hydrolysis.

Finally, we call attention to two sources of nutrients for bacteria that off-hand might seem unlikely, The water-soluble extractives in redwood include cyclitols (30 percent), tannins (25 percent), polyphenols (18 percent), polysaccharides (1.5 percent), and simple sugars (0.5 percent), all of which can be metabolized by some bacteria.[38] Indeed, one of the cyclitols, inositol, is an essential growth factor for many microorganisms. Inositol is the generic name for a mixture of hydrated stereoisomeric hexols of cyclohexane, $C_6H_6(OH)_6 \cdot 2H_2O$. Perhaps even more surprising, formulations of some wood preservatives and sealers contribute substances that can be used as sources of carbon by bacteria. Coal tar creosote, for example, is a mixture of phenolic compounds, isomers of cresol, oxycresol, and methylcresol, with lesser amounts of anthracene and naphthalene. These are examples of chemicals that at high concentration are toxic to microorganisms, but at low concentrations serve as food for nutritionally versatile organisms such as the pseudomonads. It is this remarkable adaptability that enables *Pseudomonas aeruginosa* to thrive in environments that other organisms cannot tolerate. McCulloch[39] has appropriately referred to this species as an active antagonist, for it is often found in cooling water and other places, unaccompanied by any other bacteria.

References

(1) Bungay, H. R., III. 1968. Microbial interactions in continuous culture. In *Advances in applied microbiology*, vol. 10, eds. W. W. Umbreit, and D. Perlman, New York and London: Academic Press.

(2) Chaplin, C. E. 1952. Bacterial resistance to quaternary ammonium disinfectants. *J. Bact.* 63:453.

(3) Kiser, J. S., Gale, G. O., and Kemp, G. A. 1969. Resistance to antimicrobial agents. In *Advances in applied microbiology*, vol. 11, ed. D. Perlman, New York and London: Academic Press.

(4) Lowick, J. H. B. and James, A. M. 1957. The electrokinetic properties of *Aerobacter aerogenes.* A comparison of the properties of normal and crystal violet-trained cells. *Biochem J.* 65:431.

(5) Benarde, M. A., Snow, W. B., Olivieri, V. P., and Davidson, B. 1967. Kinetics and mechanism of bacterial disinfection by chlorine dioxide. *Appl. Microbiol.* 15:257.

(6) *Standard methods for the examination of water and wastewater.* 1971. 13th ed. American Public Health Assoc., New York.

(7) Latimer, W. M. 1952. *The oxidation states of the elements and their potentials in aqueous solutions.* 2nd ed. Englewood Cliffs: Prentice-Hall.

(8) Chambers, C. W. and Clarke, N. A. 1966. Control of bacteria in nondomestic water supplies. *Advan. Appl. Microbiology* 8:105.

(9) Petterson, R. C. and Grzeskowiak, U. 1959. N-halogen compounds. I. decomposition of 1,3-dichloro-5,5-dimethylhydantoin in water at pH 9. *J. Org. Chem.* 24-1414.

(10) Petterson, R. C., Grzeskowiak, U., and Jules, L. H. 1960. N-halogen compounds. II. The N-Cl stretching bond in some N-chloramides. The structure of trichloroisocyanuric acid. *J. Org. Chem.* 25:1595.

(11) Ishii, Yoshii, Ito, Toshiko, and Kato, Shieaki. 1958. Synthesis of hydantoin derivatives and halogenation of N-halohydantoin. *Kogyo Kagaku Zasshi* 61:1254; *C. Abst.* 56:467e.

(12) Myhrstad, J. A. and Sandal, J. E. 1969. Behavior and determination of chlorine dioxide. *J. Am. Water Works Assoc.* 61: 4, 205.

(13) Bernhard, S. A. 1968. *The structure and function of enzymes.* New York: W. A. Benjamin, Inc.

(14) Buckman, S. J. and Pera, J. D. Feb. 28, 1967. *Composition containing methylene bis-thiocyanate, dispersant, and dimethylamide, and processes of inhibiting microbiological deterioration utilizing said composition.* U.S. Patent No. 3,306,811.

(15) Happold, F.C., Johnstone, K.I., Rogers, H.J., and Youatt, J.B. 1954. The isolation and characteristics of an organism oxidizing thiocyanate. *J. Gen. Microbiol.* 10:261.

(16) Wehner, D. C. and Hinz, C. F. 1971. Organic thiocyanate as industrial antimicrobial agents. chapter 45, vol. 12, *Developments in industrial microbiology.* Soc. for Ind. Microbiol., New York: Plenum.

(17) Siegel, J. R. and Rosenblatt, D. H. 1958. Formation of thiocyanate ion through neighboring group displacement reactions. *J. Am. Chem. Soc.* 80:1753.

(18) Stack, V. T. Jr. 1957. Toxicity of alpha, beta-unsaturated carbonyl compounds to microorganisms. *Ind. Eng. Chem.* 49: 913.

(19) Legator, M. June 6, 1961. *Inhibiting growth of bacteria in water flooding process.* U.S. Patent No. 2,987,475.

(20) Pelcak, E. J. and Dornbush, A. C. Sept. 29, 1959. *Process water treatment.* U.S. Patent No. 2,909,595.

(21) Sanborn, J. R. 1944. Slime-producing coliform and coliform-like bacteria. *J. Bacteriol.* 48:211.

(22) S.C.I. Monograph No. 19. 1965. *Surface activity of the microbial cell.* London: Society of Chemical Industry. and New York: Gordon and Breach Science Publishers.

(23) James, A. M. *Surface-active agents in microbiology. Ibid.*, pp. 3–23.

(24) Sykes, G. *An introduction to the use of surface-active agents. Ibid.*, pp. 159–168.

(25) Sexton, W. A. 1963. *Chemical constitution and biological activity.* 3rd ed., Princeton: Van Nostrand.

(26) McCalla, T. M. 1940. Cation adsorption by bacteria. *J. Bacteriol.* 40:23.

(27) Gratteau, J. C. Nov. 28, 1970. Potential algicides for the control of algae. *Water and sewage works,* 117:R24.

(28) Hugo, W.B. 1967. The mode of action of antibacterial agents. *J. Appl. Biol.* 30:17.

(29) Benarde, M. A., Snow, W. B., Olivieri, V. P., and Davidson, B. 1967. Kinetics and mechanism of bacterial disinfection by chlorine dioxide. *Appl. Microb.* 15:257.

(30) Pressman, B. C. 1963. The effects of guanidine and alkylguanidines on the energy transfer reactions of mitochondria. *J.*

Biol Chem. 238:409.

(31) Albert, A. 1963. Correlations between microbiological morphology and the chemistry of biocides. In *Advances in applied microbiology,* vol. 5, ed. W. W. Umbreit, New York and London: Academic Press.

(32) Conkey, J. H. 1969. Relative toxicity of biostatic agents suggested for use in the pulp and paper industry. 1968 review. *TAPPI* 52:2311.

(33) Newton, B. A. 1960. The mechanism of the bactericidal action of surface active compounds: a summary. *J. Appl. Bact.* 23:2, 345.

(34) Plunkett, E. R. 1966. *Handbook of industrial toxicology.* New York: Chemical Publishing Company.

(35) Fischer, R. and Larose, P. 1952. Factors governing the adaptation of bacteria against quaternaries. *Nature* 170:715 (London).

(36) Baechler, R. H., Bendtsen, B. A., and Roth, H. G. Apr., 1966. Preservative treatment of cooling tower slats. *Heating, Piping and Air Conditioning* 38:4, 121.

(37) Gjovik, L. R., Bendtsen, B. A., and Roth, H. G. Condition of Preservative-treated cooling tower slats after 10-year service. *Forest Products Journal* 22:4, 35.

(38) Jones, F. E. and Greenberg, A. E. 1964. Coliform bacteria in redwood storage tanks. *J. Am. Water Works Assoc.* 56:1489.

(39) McCulloch, E. C. 1945. *Disinfection and sterilization.* pp. 34, 280, Philadelphia: Lea and Febiger.

VI.

Systematic Application of Microbicides

Before attempting to eliminate the nuisance microorganisms in a cooling system it is appropriate to examine the conditions under which they live, as well as the nature of their habitats. Borrowing the terminology used for bacterial populations in lakes and ponds, those that are free-swimming in the water are termed planktonic, and those living in accumulations of mud are termed benthic. In addition to these, numerous bacteria are attached to the surfaces of piping and to the walls of the basins of cooling towers; these are termed adherent. Potter[1] has reported that Gram-positive spore-forming bacteria are often found in large numbers in the mud at the bottom of lakes and ponds, but seldom in the water, where Gram-negative rods predominate. Thus, in these natural environments two distinct populations exist.

1. THE MICROBIOLOGY OF SURFACES

The number of bacteria adhering to the surfaces of stones, or particles of sand removed from a lake or pond is far greater than the number free-swimming in the water, although the adherent bacteria are more typical of planktonic than of benthic bacteria. Meadows and Anderson[2] found bacteria, blue-green algae, and diatoms on particles of sand from both marine and fresh water. Evidently stones and suspended matter adsorb nutrients from the water, which attract microorganisms, but there are still other factors. For example, in studying the interactions between bacteria and particles of clay Stotzky[3] found that montmorillonite stimulates respiration by exchanging alkali and alkaline earth cations for hydrogen ions released during metabolism. This maintains the pH of the environment at a suitable value for the growth of microorganisms, even in buffered solutions. An antithetical surface effect is that of the adsorption of compounds on clays that have surface areas of several hundred square meters per gram. Under these circumstances the rate of biological oxidation is depressed by the adsorption phenomenon, and also by the inactivation of extracellular enzymes that are themselves adsorbed on the clay.

ZoBell[4] has investigated the influence of solid surfaces on the activity of bacteria. He showed that thin bacterial films develop on surfaces where organic material is adsorbed and concentrated from dilute solutions. These surfaces then become a focus of microbiological activity, the thickness of the films increasing as the concentration of organic material in the water increases. Bott and Brock[5] note, however, that some colonies of aquatic bacteria growing on immersed microscope slides are not adapted to sessile existence; after consuming all the food in their vicinity they die. Also, in flowing water bacteria are usually of the type in which both progeny of cell division adhere tenaciously to the substrate. In stagnant water, on the other hand, the original cells remain attached, but the daughter cells often swim away.

In discussing the mechanism whereby bacteria attach themselves to a solid surface Sharpley[6] states that they adhere more firmly to positively charged surfaces than to those negatively charged, and they also are attracted to hydrophilic surfaces more strongly than to hydrophobic. Adherent bacterial cells apparently secrete a proteinaceous cement that binds them so tightly to solid substrates that they do not wash or brush off. In rare instances certain bacteria appear to develop a set of stems or stalks by which they attach themselves to firm surfaces. Adhesion to metal surfaces is a key factor in the competition for survival between slow-growing and fast-growing bacteria. Bungay,[7] in studying microbial interactions in a continuous culture, found that a slower-growing species can persist in appreciable numbers in competition with rapidly growing organisms if the slow-growing species is continually reinoculated into the bulk liquid from the growth on a surface.

In a recirculating system the cooler surfaces of piping, heat exchangers, and surface condensers are covered by bacterial films that capture organic nutrients from the cooling water as it flows over the surface. The adherent microorganisms metabolize the nutrients, multiplying in the process, and proliferating slime. If allowed to grow unhindered, the coating continues to thicken, eventually sloughing from the metal surfaces when the shear stress of the water exceeds the shear strength of the microbiological growth. In a system receiving regular treatment with microbicides, however, the major part of such accumulations is usually hydrated dirt and other insoluble material captured on the thin, sticky bacterial film. The respiration of adherent bacteria is limited by the rate of diffusion of oxygen and by the circumstance that the cells cannot exchange oxygen equally through all their surfaces. Variations in pH and oxygen tension make slime a refuge for anaerobic microorganisms. Also, the microorganisms under this mass soon become nutritionally deprived, a condition conducive to the generation of slime.

Primary microbiological growth seldom forms on surfaces where heat is exchanged because bacterial films are immediately desiccated by heat. These accumulations can slough from cooler places, however, and then bake on heated surfaces to the detriment of heat transfer. Secondary fouling of this sort normally contains much more dirt and inorganic material than biological debris.

2. *DISPERSANTS, FLOCCULANTS, AND SURFACTANTS*

From the discussion in the preceding section it can be inferred that free-swimming microorganisms in the recirculating water in a cooling system are more accessible to toxic chemicals than are those adhering to solid surfaces, for the latter are protected to some extent by extraneous matter adsorbed, or otherwise attached to bacterial films. Furthermore, in the absence of some sort of penetrant, only the outer layers of bacteria can be reached by a chemical dissolved or dispersed in the water. Thus, no matter how toxic a biocide may be it cannot be effective unless it reaches the cell walls of all the microorganisms, however situated, within a cooling system.

The role of polyacrylates as flocculants of microorganisms in the activated sludge process for treating sanitary wastes has been investigated thoroughly,[8-10] but not much has been published on their use in the microbiological control of cooling systems. Edwards,[8] however, has proposed a mechanism for the removal of sediments of silt and clays by anionic polyelectrolytes, a process that affects the number of bacteria in the water. There is no significant removal of sediment by polymeric flocculants unless the water flowing over it contains suspended solids of the same composition. That is, some of the solid phase must be present as a suspension to cause floc to interact to disperse the latter. When a flocculant is added to the recirculating suspension, particles of floc form rapidly and settle on the sediment, at the same time continuing to grow. Then, as the flocs flow or roll across the surface, particles of sediment are incorporated into the preformed flocs because of mutual adhesion, and in due time the sediment is dispersed. In this process bacteria in the sediment are likewise dispersed into the water, increasing the number of free-swimmers. Anionic polyacrylates having molecular weights around four million invariably cause an apparent increase in bacterial counts when used to disperse accumulations of mud. In biological waste treatment a natural bio-flocculation occurs in the terminal stage of the declining growth, or death, phase (see Chapter III, page 37), initiated by natural polyelectrolytes secreted by the microorganisms themselves. Proteins, protamines, alginic acid, and some polysaccharides cover the external surfaces of the aging bacteria, causing them to flocculate. Tenney and Stumm[9] report that centrifuging a culture of slowly respiring bacteria can shear off quantities of loosely bound capsular material. They also found that cationic polyelectrolytes caused rapid flocculation at 20–200 ppm, but no flocculation up to 10 ppm; above 200 ppm microorganisms began to redisperse in the water. The flocculation of dispersed microorganisms seems to be brought about when polymers are adsorbed on the surfaces of cells forming molecular bridges between them.[10]

A few chemicals act as microbicides at high concentrations, and as surfactants at low concentrations. Examples are dodecylguanidine acid salts and quaternary ammonium salts. As surfactants, at 5–10 ppm, these compounds are useful for dislodging bacteria and other debris from solid surfaces, and algal growths from the structural members of cooling towers. Sodium dioctylsulfosuccinate is especially valuable for removing protective coatings of oily material from the surfaces of bacterial films; oil is impervious to many biocides.

3. EVALUATION OF TREATMENTS

In the great majority of cooling water systems an attempt is made to control microbiological growth by adding biocides on some sort of calendar schedule. Sometimes a system is chlorinated every other day, holding some specified residual for a certain length of time. Then, perhaps twice a month, a so-called nonoxidizing biocide (which, in effect, means something other than chlorine) is added to the system. Almost always this is wasteful and expensive, and often it is also ineffective. Even more wasteful is continuous treatment with biocides. In general, the best results are obtained by adding periodically, all at once, sufficient biocide to produce a relatively high concentration in the recirculating water, typically 30–50 ppm.

The most effective and economical treatment for a specific installation can be determined semi-quantitatively by making viable plate counts periodically, and adding biocide as required to reduce the number of bacteria in the water. Plate counts are easy to make, require a relatively modest expenditure for equipment, and no special training is needed. Results are expressed as number of living organisms per milliliter of water. By also keeping notes on operating conditions, weather, wind, biocide used and its concentration, together with a continuity of sampling, a valuable history can be compiled. Detailed directions for making viable plate counts are given in Chapter IX; a general discussion of the procedure follows.

3.1 Viable Plate Counts

The viable plate count is an empirical method for estimating the density of living bacteria in a sample of water. When carried out according to the procedure in the Standard Methods for the Examination of Water and Wastewater[(11)] the result is called a standard plate count. A simplified method is specified in Chapter IX that gives adequate results for routine microbiological control in cooling water systems.

A 0.1-ml portion of a freshly drawn sample of cooling water, or a dilution thereof, is spread uniformly over prepared agar in a covered Petrie dish, and then is incubated at 30-37 C for 48 h. The number of colonies of bacteria that develop on the plate is counted, multiplied by the dilution factor, if any, and recorded as organisms per milliliter of the original sample. A colony is a group of bacteria derived ideally by multiplication of a single organism. They are usually visible to the naked eye, although colonnies close together remain small. Isolated colonies, however, are large, especially those formed from rapidly growing genera such as the pseudomonads, which may produce colonies 5 mm in diameter in 24 h.

Agar is an inert, colloidal, algal mucilage obtained from the red seaweed, *Gelidium amansii* (Ceylon moss), and other species, that has been used for centuries as a thickener and gelling agent in Oriental cookery. The seaweed belongs to the algal subdivision *Rhodophyta* (red algae); the mucilage extracted from it is a mixture of the polysaccharides agarose and agaropectin. APHA Formula Plate Count Agar (Standard Methods Agar) contains the following ingredients, and can be purchased either dehydrated, or in prepared sterile tubes ready to use. The dehydrated formula is:

Tryptone (Pancreatic digest of casein, USP) (Peptone)	5 g
Yeast Extract	2.5 g
Dextrose (Glucose)	1 g
Agar	15 g

Glucose is the source of carbon and energy for a great variety of bacteria;

yeast extract and tryptone provide essential growth factors–vitamins, amino acids, and the like. This simple medium can support the growth of many different microorganisms.

Water is indispensable for the growth of all living organisms, and is the major component of all culture media. The dehydrated agar medium is reconstituted by adding it to cold water, then heating to 100 C. It is then transferred to a pressure cooker, or autoclave, and sterilized by moist heat at 15-18 psi for 20 min. The colloidal solution thus obtained remains fluid until it cools to about 43 C, at which temperature it sets to a stiff, semisolid, transparent gel. Just before the solution sets it is poured into individual sterile Petrie plates. As agar is a complex carbohydrate that is attacked by relatively few bacteria, it is extensively used in microbiology as a solidifying agent in preparing culture media.

By dispersing organisms over the surface of an agar plate the competition for nutrients is moderated to some degree, so that slow-growing bacteria are able to produce colonies in the same environment as those that grow rapidly. Plates are occasionally seen that contain a filmy layer of slow-growing bacteria over the entire surface of the plate with discrete colonies of rapidly growing organisms superposed. Pinpoint colonies on agar plates (see Plate 10) indicate inability to reproduce; that is, a biostatic condition has been temporarily established.

The temperature of incubation is not critical: 30 C is suitable for soil bacteria, while 37 C promotes the growth of pathogenic bacteria and organisms parasitic to warm-blooded animals. The higher temperature is frequently used as growth is faster, but plates should be incubated not less than 36 h. Twenty-four hours is too short an incubation period: slow-growing bacteria may be missed because their initial growth rate following innoculation of the culture medium is quite low. The development of bacteria on agar plates is self-limiting on account of crowding, lack of nutrients, and the local accumulation of metabolic wastes, so that there is little growth after 48 h.

Dilutions should be prepared so that a plate containing 30-300 colonies is obtained for counting. In practice, dilutions of 1 : 100 and 1 : 1000 are usually suitable, prepared by making serial dilutions of the sample with sterile physiological saline solution (0.85 percent NaCl in water). The dilutions must be shaken vigorously to ensure uniformity, for after cell division bacteria tend to cling together in groups, forming chains, or trichomes. Also, cationic surfactants such as dodecylguanidine hydrochloride and quaternary ammonium salts cause clumping of bacteria, which can lead to low plate counts.

Two obvious deficiencies of the colony count technique are that the majority of soil bacteria exist as aggregations rather than as single cells,[12] and that a single culture medium can support the growth of only a small proportion of the total population.[13] Both of these circumstances lead to low bacteria counts. Still another weakness is that counts obtained at two different dilutions are not likely to be comparable, nor are values obtained on the same sample by two different laboratories. Egdell, et al.[14] have published a detailed report of a program of cooperative testing among several laboratories in Great Britain in which they show that even with meticulous attention to details, reproducibility is not particularly good. Standard plate count agar is vulnerable to criticism because it is enriched with tryptone and yeast extract, and thus favors the growth of zymogenous (fermenting) over autochthonous (native) bacteria. The authors, however, found little to choose between a medium that contained these factors and another that did not. The effect of pH of the culture medium on colony counts was also inconclusive in some experiments, but in others media at pH 6.75 gave higher counts than those at 6.25. Still other sources of error are improperly sterilized sample containers, variations in the number of cells adhering to pipettes and other glassware, and mistakes in counting.

Some organisms, for example, *Desulfovibrio*, grow only on special media, but colony counts of *Aerobacter* and *Pseudomonas*, which normally predominate in cooling water, provide a reasonably reliable indication of the concentration of bacteria in the water. There are, however, a number of inherent weaknesses in the method that arise from the elusive nature of living organisms. In a medium inoculated with several kinds of microorganisms, only those that can grow in it will reproduce. Thus, most molds and fungi are overwhelmed by bacteria, while algae and nitrifying bacteria cannot grow at all in this medium. Species of *Aspergillus* are widely distributed in nature and their spores are fairly common in the air. Thus, these molds are occasionally seen growing on agar plates, although their effect within a cooling system is imponderable. Now and then a single plate in a set will show an adventitious growth of mold from the air in the laboratory. Those that are present in a sample of cooling water, however, will show a similar amount of growth on each of a series of replicate plates. Sharpley[6] has pointed out that the low concentrations of nutrients in most cooling waters support a population of microorganisms that is nutritionally undemanding, and that may be unable to grow on a highly nutri-

Plate 1. Smear of a yellow bacterial colony–probably *Flavobacterium*–stained with crystal violet, and photographed at 1000 X. Bacteria from cooling systems all look like this–short rods easily identified under high magnification. Daughter cells of bacteria that divide transversely, however, may remain attached to each other after fission, in which case they appear as filaments, as in this photomicrograph.

Plate 2. Tangle of algae removed from a fill slat of a cooling tower, and photographed at 200 X. This is a segmented filamentous alga. A few spores can be seen, along with a good deal of wind-borne dirt trapped on the sticky surfaces of the filaments.

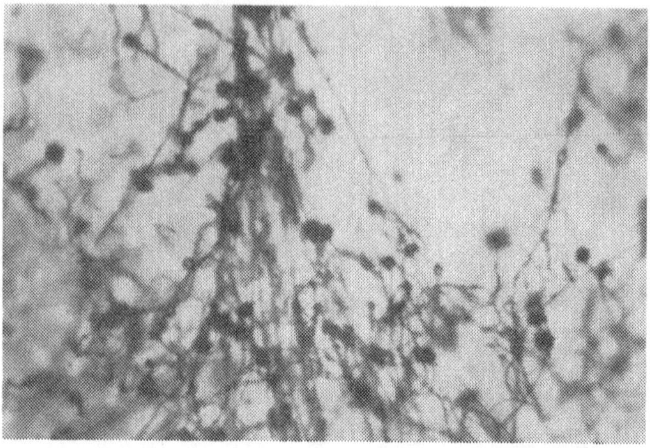

Plate 3. Thallus of filamentous fungus, or mold, showing a mycelium of thread-like hyphae with numerous spores attached. This photomicrograph was prepared by suspending the mold in water, then photographing it at 100 X.

Plate 4. *Protozoa* residing in mud from the bottom of a cooling tower basin. These forms do not normally appear within the cooling system itself, although now and then they are found in sludge in compressor water jackets. These organisms are indigenous to ponds and similar bodies of stagnant water. In the center of the photomicrograph is a bell-shaped ciliate, *Vorticella*. Its vibrating cilia are visible around the mouth of the bell, which is supported on a long flexible stalk, here encrusted with flakes of ferrous sulfide. Immediately above the bell is a disc-shaped organism, probably a yeast cell, or a protozoan cyst. To the right is another common ciliate, the *Paramecium*, and immediately to the left is a rotifer, *Philodina*. Several of the latter are more clearly shown in Plate 5. The magnification in Plate 4 is 200 X.

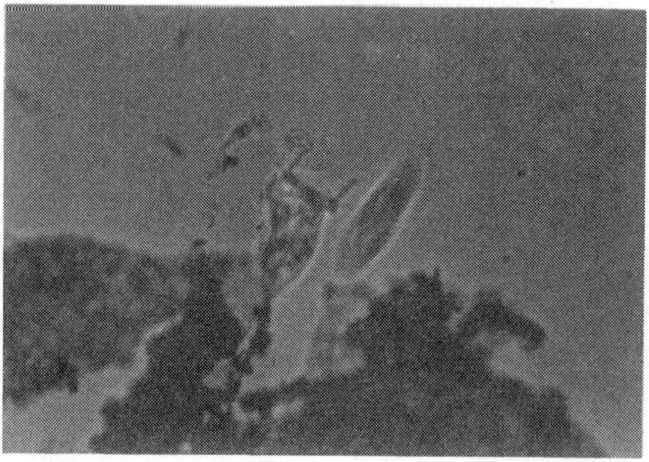

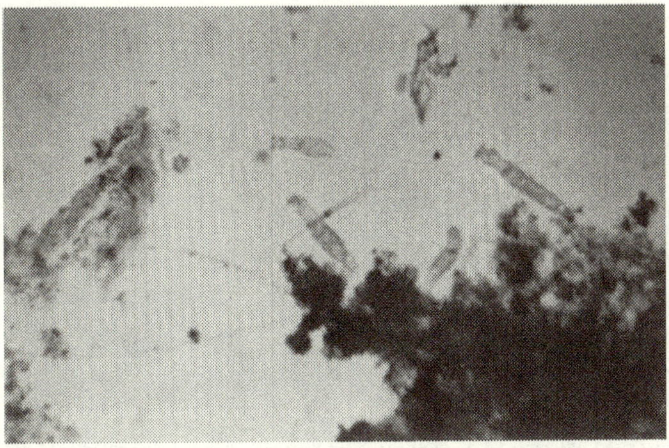

Plate 5. Rotifers attached to sludge from a cooling tower basin. These organisms have a forked foot, clearly visible on the one furthest to the right, which has just detached itself from the mass of hydrated soil. These organisms forage by waving around in the water, while sweeping bacteria and other organic particles into their mouths by means of cilia, which because of their particular motion resemble rotating wheels. A few strands of decomposing filamentous algae are visible in the lower left quadrant of the photomicrograph. Magnification is 100 X.

Plate 6. A roundworm residing in mud from a cooling tower basin, photographed at 100 X. Worms are rarely seen in basin mud—probably because they are nutritionally more demanding than the other simpler animal species, and they cannot survive long in this barren environment.

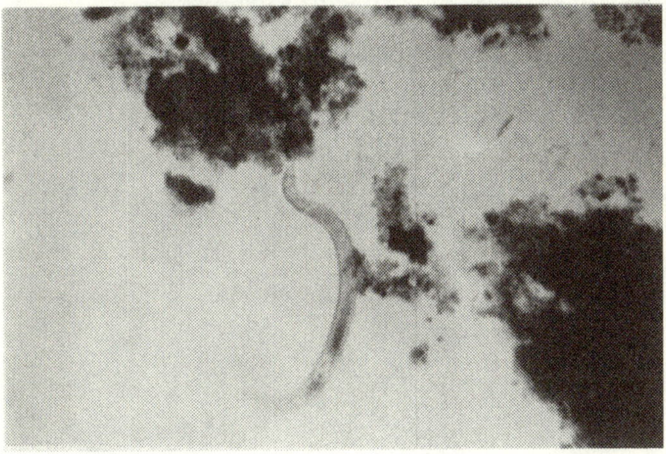

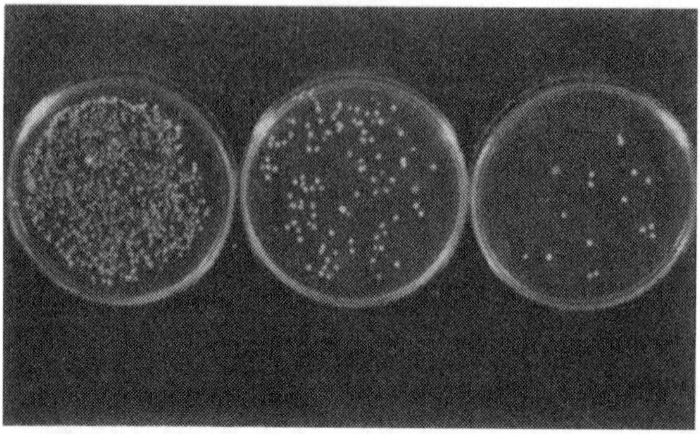

Plate 7. Colonies of bacteria growing on nutrient agar at three successive dilutions: 1:100, 1:1000, 1:10,000, left to right. Note the smaller colonies on the 1:100 plate caused by crowding. The number of colonies on each plate from left to right are 1380, 146, and 18; the average plate count is 1,500,000 bacteria per milliliter. The colonies in the photograph are viewed from above through the bottoms of inverted petrie plates.

Plate 8. Colonies of *Aspergillus niger* growing on an agar plate. The white corona is a thick tangle of segmented yellowish-green hyphae bearing many spherical conidiophores. Conidiophores, separated from the hyphae, lie around and under the black centers, which are conidiophores that have burst, liberating black thistle-like efflorescences of spores.

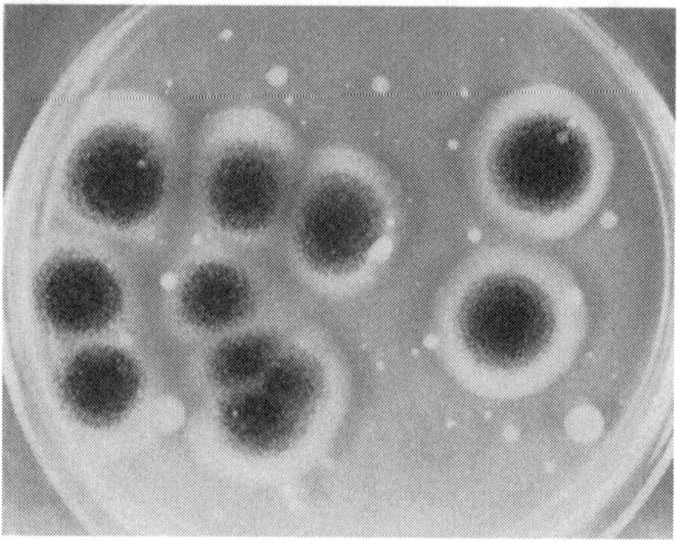

Plate 9. Bacteria in a sample of cooling water, stained with crystal violet, and photographed at 400 X. Note clumping of the bacteria, a phenomenon that makes the accuracy of plate counts uncertain.

Plate 10. Pin-point colonies on an agar plate, streaked two days after treating a cooling system with chlorine dioxide at 6 ppm. These bacteria are moribund, for although this plate indicates 3,970,000 bacteria per milliliter, in three days it had fallen to 250,000 per milliliter, indicating that 94 percent of the organisms on this plate had expired in the interim. The colonies are viewed from above into the open petrie plate.

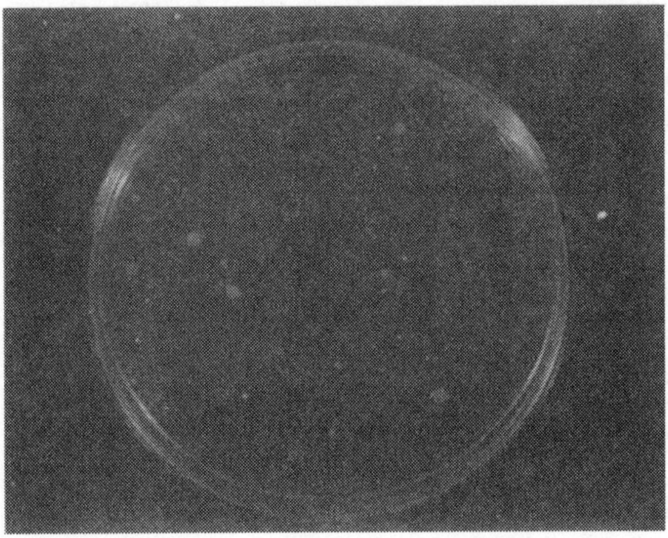

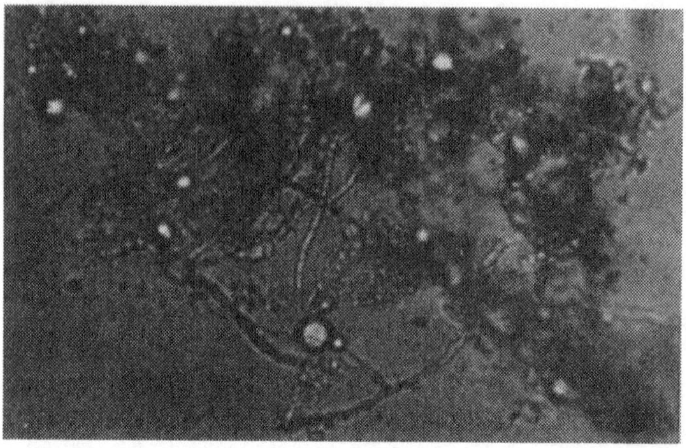

Plate 11. Filamentous bacteria (probably *Sphaerotilis natans*) growing in a mat on the surface of the water in the basin of a cooling tower following contamination of the system by mixed xylenes. Photographed at 400 X. Note individual droplets of xylene, as well as xylene-water emulsion. Once established, these bacteria are extremely troublesome to eradicate.

Plate 12. *Aspergillus*-like mold infesting a cooling system after a prolonged leak of monoethanolamine, photographed at 100 X. The mold is visible as suspended flocs in the recirculating water. It plugs screens rapidly, and combined with dirt, fouls heat exchangers and compressor jackets. The spores are extremely resistant to most biocides.

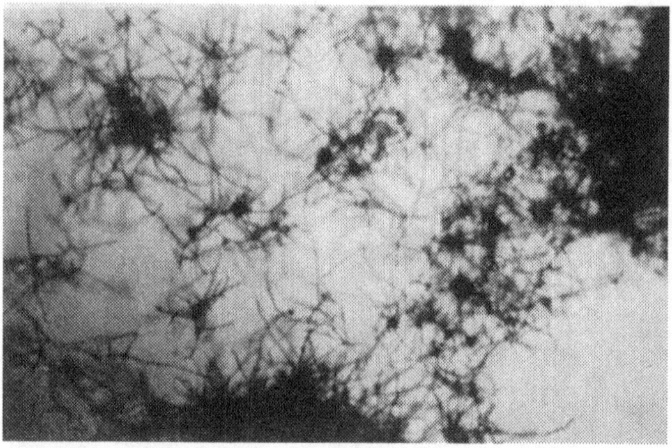

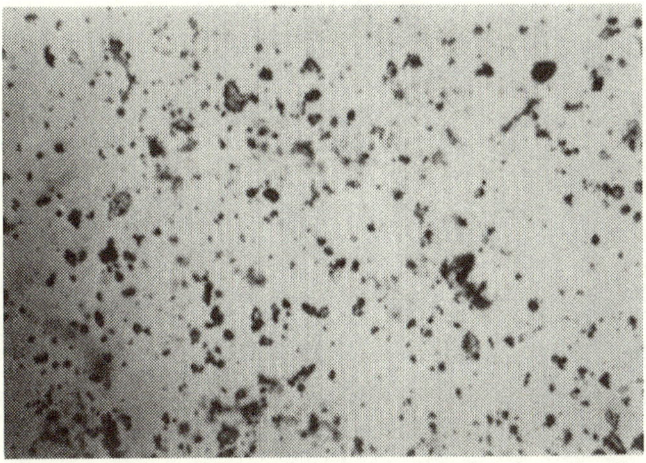

Plate 13. Particles of dirt removed from the intake louvers of an operating cooling tower, and photographed at 200 X. When this is swept into the tower it becomes hydrated and appears as gelatinous mud in places in the cooling system where the flow of the water is slow. Note the irregular shape of the particles, which conclusively distinguishes them from microorganisms; the latter are invariably symmetrical.

Plate 14. Particles of rust magnified 100 X. Here again, the particles are irregularly shaped, and obviously not microorganisms.

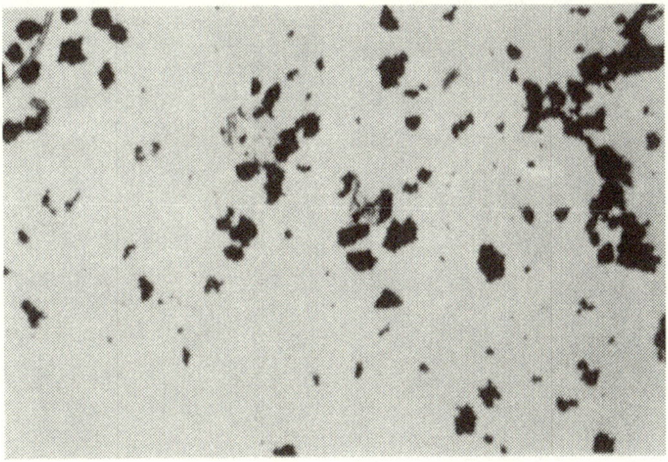

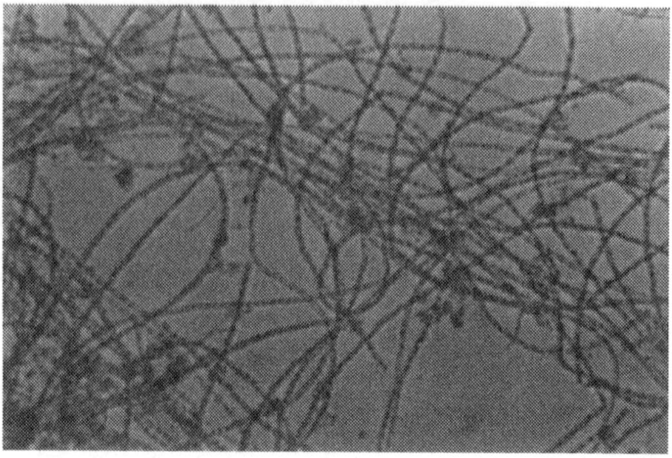

Plate 15. Filamentous algae, photographed at 100 X. Practically anything that is slimy or gelatinous is described by practitioners in the water-treating field as algae, or "dead algae." Microscopic examination provides unequivocal proof of the nature of various slimes and miscellaneous debris; the unaided eye does not.

Plate 16. Ferric hydroxide slime, photographed at 200 X. Chemical slimes of this sort are often erroneously labeled microbiological.

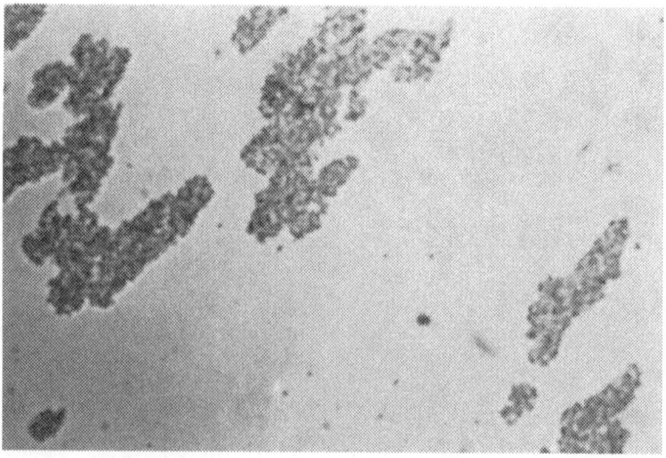

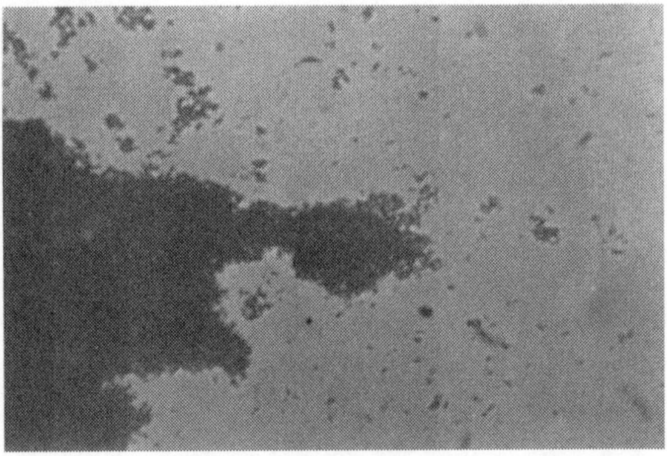

Plate 17. Bacterial slime, fixed, stained with crystal violet, and photographed at 400 X. Note the small rods around the edges of the stained mass. Their regular shape, characteristic appearance, and size all suggest bacteria. Slime layers vary in thickness and consistency, but as they are not dense, they are usually invisible under the microscope. Also, capsular material has little affinity for dyes, so that when fixed and stained the bacteria within an accumulation of slime appear as a mass of rods, as in this photomicrograph.

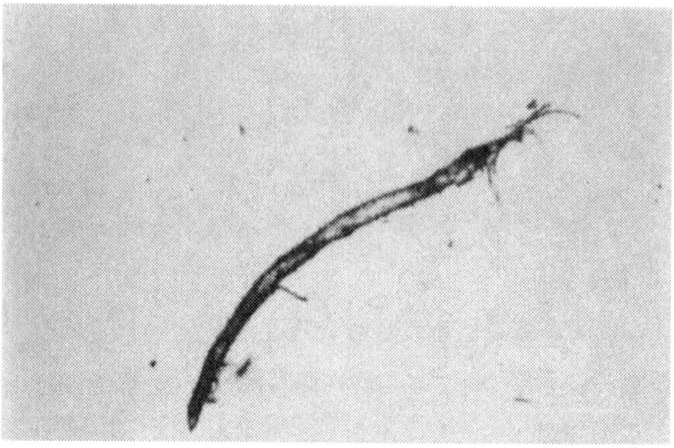

Plate 18. Redwood fiber, photographed at 100 X. These are prevalent in deposits from cooling systems with redwood cooling towers, and can lead to erroneous interpretations of the loss-on-ignition determination, used by some as an indication of the presence of microbiological debris.

tious substrate. Finally, because of the high dilutions required, unusual ratios of genera may be observed.

Viable plate counts less than 10,000/ml are seldom obtained, even immediately after treatment with biocide, because open cooling systems are continually being inoculated. Counts of 100,000/ml-500,000/ml indicate a biologically clean system; when counts rise above one million treatment with a toxicant is advisable. With the exception of *Aerobacter aerogenes*, coliform bacteria are rare in cooling water, although *Escherichia coli* has been reported occasionally, presumably introduced by birds. The presence of coliform bacteria can be confirmed by streaking a plate prepared with eosin-methylene blue (EMB) agar, which contains peptone and lactose in addition to the two dyes. The fermentation of lactose produces an acid that causes colonies of *Escherichia coli* to assume a deep blue-black color with a metallic luster by reflected light. Colonies of *Aerobacter aerogenes* are pale pink, or lavender, without sheen. There is seldom any reason to be concerned with these species.

The American Petroleum Institute has published a procedure for detecting the presence of sulfate-reducing bacteria.[15] Sulfate API broth can be purchased in prepared vials containing 9 ml of the medium. Each container is sealed with a rubber septum through which the sample is introduced with a sterile syringe. Included in the culture medium is an iron salt that reacts with hydrogen sulfide generated by the bacteria to form a black precipitate of ferrous sulfide. The sample of water is added without dilution to avoid killing anaerobes with air. Details of the procedure are given in Chapter IX; the complete formula of the medium is as follows:

Sodium lactate (50-60 percent)	5.0 ml
Yeast extract	1.0 g
Ascorbic acid	0.1 mg
$MgSO_4 \cdot 7H_2O$	0.2 mg
K_2HPO_4	0.2 mg
$Fe(SO_4)_2(NH_4)_2SO_4 \cdot 6H_2O$	0.1 mg
NaCl	10.0 g
Agar	20.0 g
Distilled water (pH 7.0-7.5)	1000 ml

The large expense of microbiological control can be greatly reduced by using plate counts to show when biocide is needed, and whether a treatment is effective. Carrying this concept still further, Chambers and Clark[16] suggest that plate counts taken at intervals of one to several hours, and correlated with simultaneous determinations of residual toxicant show, under actual operating conditions, the minimum effective concentration of microbicide.

3.2 Membrane Filter Counts

The Millipore total-count sampler consists of a plastic tab supporting a Millipore membrane filter bonded to an absorbent pad containing dehydrated Tryptone-Glucose-Yeast Extract nutrient medium. When the sampler is immersed in a portion of diluted sample the absorbent pad quickly draws one milliliter of water through the Millipore filter, and into the pad. Bacteria are retained on the gridded filter, as they are too large to pass through the pores, which are 0.45 millimicrons in diameter. Water drawn into the pad immediately hydrates the nutrient medium, which then diffuses through the pores to the surface of the filter where it provides nutrient for the retained bacteria. The sampler is incubated, and the resulting colonies are counted, as in the viable plate count procedure. Much less working time is required for the Millipore total count, the simple procedure demands little of the user, and some of the difficulties of the spread-plate method (uneven distribution, confluent growth, contaminated plates, etc.) are avoided.

When using the Millipore total-count sampler,[17] water samples are diluted as in the viable plate count method (see page 113), except that the final dilution is made in the outer sleeve of the sampler by adding to it 2.0 ml of the preceding dilution followed by 18.0 ml of sterile saline solution. After mixing, the plastic tab supporting the membrane filter and absorbent pad is immersed in the dilution for 30 s, then removed. The dilution is discarded, the sampler is returned to the sleeve, and the entire assembly is incubated for 48 h. The Millipore filter is grey, but after wetting turns black and provides a contrasting background for the white or clear colonies of bacteria. Colonies are best counted under direct illumination magnified five- to ten-fold. Counts obtained using the Millipore total-count sampler are about 30 percent lower than those obtained on agar plates.

3.3 Direct Counts

This method of enumerating bacteria measures all microorganisms, including dead ones, that are present in a sample of water. In addition to bacteria other particulates are likely to be counted as well, so this method invariably yields much higher numbers than does a viable plate count, which of course, measures only those organisms that are alive and can grow on a particular medium. Furthermore, the viable count rests on the assumption that each colony develops from a single cell, whereas it is entirely possible, or even probable, that many of them arise from aggregates containing numerous microorganisms. The clumping of bacteria makes any method of counting extremely uncertain, especially in large systems, where the size of the sample taken is minute compared to the total volume of water in the system.

Direct counting is difficult, tedious, and time-consuming under the best of conditions, so its routine use can hardly be justified for cooling waters. Special counting chambers are used containing ruled grids to facilitate counting under the microscope; the addition of a small amount of crystal violet improves viewing of motile bacteria. The interested reader should refer to a book by Collins[18] for further information on direct counting methods.

3.4 Bioluminescence Counting

A promising new method for indirectly measuring the total concentration of living microorganisms in water is based on the classic bioluminescent reaction of adenosine triphosphate (ATP) with the luciferin-luciferase enzyme system to form adenosine monophosphate, simultaneously releasing a photon of light.[19] The intensity of the latter is measured in a liquid scintillation spectrometer.[20,21] The coenzyme adenosine triphosphate is present at a relatively uniform concentration in the cells of all living organisms, where it is formed by the oxidative phosphorylation of adenosine diphosphate during the metabolic transfer of hydrogen atoms to oxygen from the substrate. The energy of the latter reaction couples inorganic phosphate to adenosine diphosphate yielding ATP, which contains two energetic oxygen-phosphorus bonds that can oxidize luciferin with the evolution of light.

Reid and Leese[22] formulate the synthesis of ATP as shown in the following equation.

$$\underset{\textit{adenosine diphosphate}}{\text{ADP}} + \underset{\textit{inorganic phosphate}}{i\text{PO}_4} + \text{energy} = \underset{\textit{adenosine triphosphate}}{\text{ATP}} + \text{H}_2\text{O} \tag{6.1}$$

The energy to drive the above reaction is derived from a series of steps wherein some suitable substrate is oxidized by oxygen, the electrons being transported through a chain of carriers in the mitochondria from the substrate to oxygen, producing in the process three molecules of ATP per pair of electrons. Mitochondria in plant and animal cells are organelles that contain respiratory enzymes. Peter Mitchell has formulated a chemiosmotic hypothesis to explain how such an energetic molecule as adenosine triphosphate can be synthesized in living cells. He postulates that the electron carriers are arranged in a regular pattern across the inner mitochondrial membrane, giving rise to an electrochemical transmembrane gradient that he calls proticity. This gradient, it is assumed, drives the ATP synthesis represented by Eq. (6.1). Although bacteria do not contain mitochondria, they perform the same chemical reactions with respiratory enzymes contained in their cytoplasmic membranes, and thus ATP is present to give the luciferin reaction.

ATP is present only in living microorganisms, for when it is released from a living cell it is immediately inactivated by adenosine triphosphatase. Therefore, in common with the viable plate count, only living cells are measured, and the two methods correlate fairly well. The sample of water to be counted is first boiled in a tris(hydroxymethyl)aminomethane buffer solution at pH 7.5 to lyse the living cells, releasing ATP, and also to deactivate the adenosine triphosphatase enzyme, which would otherwise rapidly degrade the released ATP. The lysed solution is then centrifuged, and a small portion of the supernatant liquid is mixed with a measured volume of luciferin firefly extract in a scintillation vial, then measured in a photometer.

As yet, this method has not been extensively used for measuring microorganisms in cooling water, as the equipment is expensive and a certain amount of skill is required to obtain consistent results; also there are some difficulties in calibration and the method is not specific. The interested reader is referred to the three references already given[19-21] for further details.

3.5 Age of Samples

Plate counts are significant only if the sample of water is fresh; the importance of this is shown in Tables 6.1 and 6.2.

Notice that in A the plate counts decreased about 90 percent, while in C the counts nearly doubled after being stored for five days. In all three samples the pH decreased significantly, probably because of carbon dioxide generated in the respiration of the bacteria.

TABLE 6.1

Effect of Storage on pH and Plate Count

		Fresh Sample		*Sample Held 5 Days*	
System	Total Capacity (gal)	pH	Plate Count	pH	Plate Count
A	1,725,000	7.64	1,400,000	7.11	170,000
B	750,000	7.99	300,000	7.47	200,000
C	425,000	6.56	800,000	6.31	1,500,000

The samples listed in Table 6.1 were taken and stored in sterile polyethylene bottles. Potter[23] found that, in general, bacterial populations multiply in natural water samples stored in glass. Her results show that unbuffered solutions above pH 6 all tend to increase to pH 8.0-8.4, the bicarbonate buffer range, undoubtedly as a result of small amounts of alkali dissolved from the glass reacting with carbon dioxide generated by the bacteria.

The results in Table 6.2 were obtained from samples stored in glass culture tubes. The first three sets of these were kept at room temperature; the fourth set was refrigerated at 2 C. The unrefrigerated samples all decreased in plate count, but the refrigerated samples remained essentially constant. There was no significant change in pH of any of the samples, which were initially at about pH 6.8, and after four days were at pH 7.0.

TABLE 6.2

Effect of Aging on Plate Count

Day	*Sample I*	*Sample II*	*Sample III*	*Sample IV**
0	640,000	960,000	620,000	750,000
1	700,000	760,000	720,000	720,000
2	260,000	450,000	400,000	730,000
3	120,000	190,000	330,000	760,000
4	250,000	130,000	330,000	850,000

* Refrigerated

The numbers of bacteria in samples of cooling water change rapidly with time as the concentrations of oxygen and nutrients diminish, and toxic metabolic wastes accumulate. Also, the rate of death increases with age; the more vulnerable die first; the more resistant persist.

3.6 Microbiological Analyses

Water service companies sometimes provide their clients with reports of microbiological analyses of cooling water. These are interesting to look at, but are of no particular value in operating a cooling system. Typically, a total count as determined by the standard procedure[11] is given, with further classification of "spore-formers," "nonspore-formers," "corrosives," "slime-formers," and others.

Spore formation among bacteria is essentially confined to *Bacillus* and *Clostridium*, neither of which is often seen in cooling water. There appears to be a connection between a low level of nutrients and the formation of spores by these two genera, but Salle[24] points out that artificial methods of inducing spore formation, such as extremes of heat and cold, or special chemicals, are laboratory phenomena that are unlikely to occur in nature. Indeed, Foster[25] considers that these unnatural conditions should be classified as pathological, being completely foreign to the normal environment.

Spore-formers are measured by pasteurizing the water sample, or a culture thereof, at 80 C for 5–10 min, or sometimes at 100 C for 15–30 min. This kills all vegetative cells so only the resistant endospores give rise to colonies in a subsequent plate count. Note that this procedure does not really measure total spore-formers, but only those that happen to be in the endospore state when the sample is pasteurized. Spore-formers in the vegetative state are killed along with the nonspore-formers.

Sulfate-reducing bacteria, in the unlikely event that any are found in a stored sample, are sometimes reported as corrosives.

A further attempt is made to classify microorganisms by the appearance of colonies on an agar plate. Colonies that show a characteristic moist, glistening sheen are called "mucoids." This is not a taxonomic term, but describes an appearance associated with the formation of a capsule, so these organisms are often reported as slime-formers. It should be noted, however, that a number of soil bacteria form capsules only under special circumstances, one of which is the presence of sugars. Thus, bacteria that form mucoid colonies on standard plate count agar might not be inclined to do this when in cooling water. In other words, what a species does on an agar plate is not necessarily what it does in its natural state.[26] Yellow colonies are reported as *Flavobacterium*, a generic name, while irregular or wrinkled colonies are classed as *Bacillus* (one of the spore-formers).

When samples are held for several days all anaerobes die, fast-growing species crowd out the slow-growing, and if a microbicide is present when the sample is taken, results are grossly affected. The delay occasioned in shipping samples to distant laboratories leads to questionable results, as only information from fresh samples is reliable in microbiological analysis.

4. THE EFFECT OF BLOWDOWN RATE ON MICROORGANISMS

Several years ago a program was initiated for treating the water in some 20 cooling systems at a petroleum refinery in the San Francisco Bay Area. For one reason or another these systems were not all operated at the same cycles of concentration, and it soon became evident that, in a general way, the higher the cycles of concentration the more biocide required to keep plate counts less than 10^6/ml. It was apparent, however, that if the biocide requirement and cycles of concentration were indeed related, a large number of plate counts would be needed for a more-or-less statistical approach to discover that relationship.

Accordingly, three of the larger cooling systems were selected (designated here as A, B, and C) that could be operated continuously at fixed cycles of concentration. A schedule was established for making bacteria counts once a week, then, whenever the counts rose to 10^6/ml or more, a biocide was added. In light of subsequent experience this method could have been improved, but at any rate it was consistently followed for 136 weeks. Data obtained in the three systems are summarized in Table 6.3.

Although systems A and B operated at virtually the same cycles of concentration, A on the average, ran two weeks longer than B between additions of biocide. Obviously, cycles alone do not explain this difference. If, however, we compare the frequency of biocide addition with the ratios of blowdown rates to total volumes of the three systems, a relationship emerges.

If b/V is plotted against frequency of biocide additions on semilogarithmic graph paper, a straight line is obtained, as shown in Fig. 6.1. The significance of b/V has been discussed in a previous publication[27] so it is

TABLE 6.3

Operating Variables and Plate Counts

System	*Volume* (gal)	*Average Make up* (gph)	*Average Blowdown* (gph)	*Avg. Cycles*	*No. of Plate Counts* 10^6 +	*Avg. weeks per biocide added*
A	727,000	67,000	14,900	4.5	21	6.5
B	424,000	27,000	5,900	4.6	30	4.5
C	1,740,000	73,000	7,300	10.0	59	2.3

TABLE 6.4

Blowdown/Volume Ratios

System	*b/V*	*Weeks Between Biocide Additions*
A	.020	6.5
B	.014	4.5
C	.0042	2.3

merely pointed out here that the depletion of a chemical in a cooling system is described by the following equation:

$$\log_e(c/c_0) = -\frac{b}{V}(t - t_0) \qquad (6.2)$$

or

$$c = c_0 e^{-b(t-t_0)/V} \qquad (6.3)$$

where

c_0 = concentration of chemical at time, t_0 (ppm)
c = concentration of chemical at time, t (ppm)
b = blowdown rate (gph)
V = total volume of water in system (gal)

Eq. (6.3) is exactly analogous to the expression for radioactive decay, with b/V corresponding to the radioactive constant. Using this equation the time required for a given treating chemical to be depleted to one-half its original concentration can be calculated, which time is analogous to the half-life of a radioactive element. Thus,

$$\log_e c - \log_e c_0 = -bt/V$$

$$\log(c_0/2) - \log_e c_0 = -bt/V$$

$$\log_e(2) = \frac{b}{V}\, t_{1/2}$$

$$t_{1/2} = \frac{0.693}{b/V} \qquad (6.4)$$

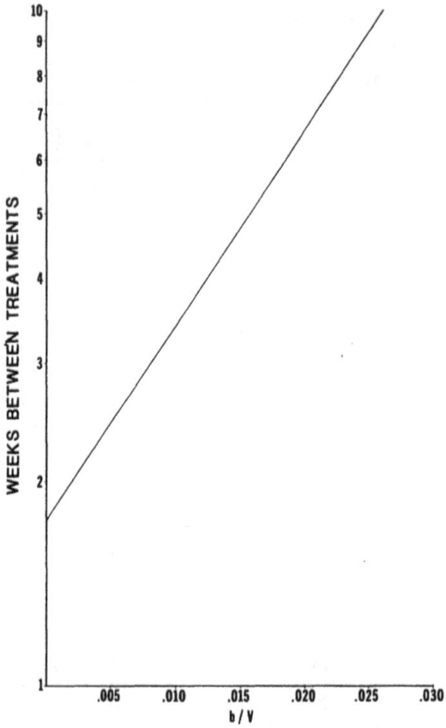

Fig. 6.1 Relationship between blowdown/volume and biocide consumption.

Table 6.5 lists the half-depletion times at various cycles of concentration for the three cooling systems under discussion.

If 20 ppm of some chemical, e.g., chromate, is added to system C operating at 10 cycles of concentration, it will take 165 h for the concentration of chromate to fall to 10 ppm, another 165 h to reach 5 ppm, and so on. Thus, system C is, in effect, a large stagnant reservoir to which great numbers of bacteria are introduced daily in wind-borne dirt. Most of these are retained in the system, not because of the high cycles of concentration, but because of the small blowdown-volume ratio in this system. In Table 6.6 the b/V values are compared in the three systems at various cycles of concentration. Referring back to Fig. 6.1 with these values an estimate can be made of how often it will be necessary to add biocide to these systems at any particular number of cycles of concentration. For example, if system B is operated at 6 cycles it would be expected that biocide would be added about once every three and one-half weeks to keep the viable plate count less than one million bacteria per milliliter.

TABLE 6.5

Half-Depletion Times

	$t_{1/2}$ hours		
Cycles	*A*	*B*	*C*
3	19	28	36
4	29	41	53
5	38	53	73
6	49	69	91
7	58	83	110
8	69	96	128
9	77	110	144
10	87	124	165

TABLE 6.6

Blowdown/Volume Ratios

	b/V		
Cycles	*A*	*B*	*C*
3	.036	.025	.019
4	.024	.017	.013
5	.018	.013	.0095
6	.014	.010	.0076
7	.012	.0084	.0063
8	.010	.0072	.0054
9	.0090	.0063	.0048
10	.0080	.0056	.0042

Before leaving this subject it would be desirable to find, if possible, the optimum cycles of concentration at which to operate a cooling system—that is, at what point the cost of inhibitors and biocides levels out. Assume for illustration that in system B 40 ppm of an inhibitor formulation that costs \$0.40/lb is maintained continuously, and that a biocide costing \$1.50/lb is slug-fed at a level of 40 ppm at the frequency indicated in Fig. 6.1. The cost of the inhibitor is calculated from the formula:

$$\text{Inhibitor, \$/day} = \frac{40 \times 200 \times 0.40 \times b}{10^6}$$

The average cost of biocide is

$$\text{Biocide, \$/day} = \frac{40 \times 424{,}000 \times 8.33 \times 1.50}{10^6 \times \text{days}}$$

where days is the number of weeks per addition of biocide (from Fig. 6.1) multiplied by seven. From these data the entries in Table 6.7 can be calculated.

TABLE 6.7

Cost of Treating System B

Cycles	b/V	*Inhibitor* (\$/day)	*Biocide* (weeks/addition)	*Biocide* (\$/day)	*Total cost* (\$/day)
3	.025	34.16	9.0	3.36	37.52
4	.017	22.77	5.3	5.71	28.48
5	.013	17.08	4.1	7.38	24.46
6	.010	13.66	3.4	8.90	22.56
7	.0084	11.39	3.0	10.09	21.48
8	.0072	9.76	2.8	10.81	20.57
9	.0063	8.54	2.6	11.64	20.18
10	.0056	7.59	2.5	12.11	19.70

From Table 6.7 it can be concluded that little benefit is realized by operating system B at more than 7 or 8 cycles of concentration, in which range total chemical cost levels out at about 20 dollars per day.

It is emphasized that the foregoing conclusions are based on average performance over a long period of time. The behavior of living organisms is not notably predictable, but regardless of that, it is certain that the ratio b/V has a great influence on the effectiveness of any program for controlling the growth of microorganisms in recirculating cooling systems.

Finally, it is worth noting that cycles of concentration depend upon heat load. As an illustration of this, suppose that system C is evaporating 66,150 gph, and blowing down at 9450 gph. The make-up rate is the sum of these two values, i.e., 75,600 gph, and the number of cycles of concentration is 75,600/9450, or 8.0. Now if the heat load is increased so that 100,000 gph are being evaporated, then with the blowdown unchanged, the make-up rate will be 109,450 gph, and the cycles of concentration will be 109,450/9450, or 11.6. Under both operating conditions b/V remains constant at .0054. The emphasis in water-treating practice on cycles of concentration, while disregarding b/V, obscures to a great extent the interpretation of the performance of an evaporative cooling system.

5. *SYSTEMATIC DOSAGE AND TREATMENT*

In order to keep cooling systems functioning it is essential that the growth of microorganisms, a matter discussed at some length in Chapter V, be kept under control. Soil bacteria in dust are blown into open cooling systems where they proliferate in the warm water. One gram of average soil may contain 50–100 million bacteria, mostly of the *Pseudomonas* genus;[28] rich garden soil may harbor a billion microbes per gram; old grassland, which has low pH, may contain only 10–20 million bacteria per gram. A 500,000-gal cooling system takes in about 500 lb of dirt daily, but in dry, windy weather the amount may increase many-fold. This continuous supply of bacteria and other microbes makes control a continuous problem.

5.1 Gaining Microbiological Control

In this section, a program is described for establishing microbiological control of a fouled cooling system, using viable plate counts as a measure of biological cleanliness. Plate counts, at the outset, should be made twice a week and the results correlated with the appearance of the tower wood. Fill slats and internal vertical members should be examined for green algae, and the walls of the basin just under the surface of the water should be checked for a film of slime. There is some evidence that growths of algae secrete nutrients and growth factors suitable for large bacterial populations.

When starting with a dirty system with clogged screens, debris, and slime and algal growths, it is difficult to gain control, but it is also expensive and ineffective to add chemicals haphazardly. No bactericide can kill bacteria unless it contacts them, and this is virtually impossible in a dirty system. Sludge in the basin is a breeding place for anaerobes and a few higher organisms, and screens clogged with negatively charged debris remove cationic biocides from the water before anything is accomplished. As a first step, therefore, the screens and basin sould be cleaned, after which the system should be blown down thoroughly.

Assuming now that there is a growth of algae in the cooling tower and some residual suspended matter in the recirculating water, an appropriate microbicide must be selected for the next step. Chlorine has little penetrating power, and is rapidly inactivated by masses of organic material. Quaternary ammonium salts and other cationic surfactants are markedly reduced in effectiveness by organic material because of the profusion of negative charges on protein molecules. Similarly, quaternary compounds are adsorbed on particulate matter leaving the solution deficient in biocide so that bacteria, particularly *Pseudomonas*, not only survive, but thrive in their presence.

Two biocides that function well in the presence of organic material are methylene bisthiocyanate and chlorophenol. Either will remove most accumulations of algae in one treatment at 3-4 ppm active concentration. Methylene bisthiocyanate hydrolyzes rapidly above pH 8, so it is not suitable for systems on nonchromate treatments, which with few exceptions, are alkaline. Chlorophenols are active biocides throughout the pH range of 5.5-8.5.

After the cooling tower is free from algae it is desirable to use a toxicant that gives some protection to the wood, in addition to controlling the growth of algae and bacteria; dodecylguanidine fulfills these requirements well. It bonds to the negative carboxylate sites in cellulose, providing protection against both algae and fungi, and it is an efficient microbicide as well. This chemical should not be added directly to a dirty system, however, as it forms electrostatic bonds with organic and inorganic matter, including Fe_2O_3, $Ca_3(PO_4)_2$, and SiO_2. In some circumstances it is precipitated completely and rendered innocuous to microbes. When dodecylguanidine is first used a considerable proportion of it is adsorbed on the wood and suspended material, so plate counts do not remain low very long. The duration of effect increases with use, however, as shown in Table 6.8.

The data in Table 6.8 were collected in the process of cleaning up a 300,000-gal recirculating system associated with an ammonia manufacturing plant. The fill section of the redwood tower was covered by a growth

TABLE 6.8

Program for Microbiological Control

STEP I		STEP II		STEP III	
Add 10% Methylene bisthiocyanate (30 ppm)		*Dodecylguanidine hydrochloride* (40 ppm)		*Dodecylguanidine hydrochloride* (30 ppm)	
Day	Plate Count	Day	Plate Count	Day	Plate Count
0	1,500,000	0	2,000,000	0	1,200,000
1	10,000	2	57,000	4	40,000
5	43,000	5	23,000	11	85,000
15	215,000	7	88,000	13	110,000
18	2,000,000	10	155,000	18	200,000
(Proceed with Step II)		13	560,000	21	600,000
		15	1,200,000	27	670,000
		(Proceed with Step III)		32	800,000
				34	1,700,000

of green algae, and a thin layer of slime covered the walls of the basin; the recirculating water was turbid with dirt and debris. For the preceding six months the system had been treated daily with acrolein for a 6-h period at a residual of 0.2-0.8 ppm, with a supplemental weekly addition of 50 ppm of 1,3-dichloro-5,5-dimethylhydantoin. The water treatment is a standard chromate-zinc-polyphosphate formulation at pH 6.0-6.5. The system was blown down to a total hardness of 150 ppm $CaCO_3$, and a plate count of 1.5×10^6 bacteria per milliliter was obtained. A 75-lb portion of a 10 percent solution of methylene bisthiocyanate was added to the basin; after 24 h the plate count was 1.0×10^4 bacteria per milliliter. The algal growth was gone from the wood, but the water was quite dirty. After three days the water began to clear, and within a week it was bright and free of suspended solids. The remains of the algae on partially wetted wooden members had a brown, powdery appearance; there was no sloughing of sticky masses.

By the eighteenth day the plate count was 2.0×10^6 bacteria per milliliter, but the water remained clear with no visible slime or algae in the tower. Next, a 100-lb portion of a 35 percent solution of dodecylguanidine hydrochloride was added with results shown under Step II in Table 6.8. This first application held plate counts down for about two weeks, then each subsequent addition of the biocide kept the system biologically clean for about a month. It is significant that although the water in this system always contains 50–200 ppm of ammonia because of small leaks, the system, when under control, shows the same rate of increase of bacteria as other cooling systems in the same industrial area that contain no ammonia.

5.2 Routine Treatment with Nonoxidizing Microbicides

Once a system is under control the operator has a large number of products from which to choose. Adaptation is a common phenomenon with quaternary ammonium salts, especially below pH 7. They form insoluble precipitates with various anions, including chromate,[29] and with anionic wetting agents such as sodium dioctylsulfosuccinate. Whatever biocide is used, care must be taken that it is compatible with other chemicals in the system. For example, if lignin is used to control corrosion and fouling, chlorine should not be used, for it oxidizes lignin to aldehydes and acids, destroying completely the already meager corrosion protection of this chemical. Similarly, ferrocyanide is oxidized to ferricyanide, which has no effect in preventing corrosion. Carbamates cannot be used as biocides in systems treated with chromate as the latter is reduced to chromic ion. The nitrogen atom in polyacrylamide readily adds chlorine and the product formed sometimes precipitates. In all of these examples the effect of essential water-treating chemicals is destroyed; in the last instance chlorine demand is also excessive.

In general, much better results are obtained by adding large amounts of toxicants infrequently than by using smaller amounts often. According to Yost,[30] slug-feeding and alternating biocides discourages the development of resistant mutants, or strains, as two simultaneous mutations would be necessary for a bacterium to become resistant to two different biocides. A 30–40 ppm concentration of most commercial biocides gives satisfactory control. Low temperature slows the growth of microorganisms, so other things being equal, it may be possible to decrease the concentration of biocide in winter. Unfortunately, psychrophilic bacteria tend to produce more slime when it is cold than they do in warm weather.

The performance of biocides varies from time to time, as well as from system to system, but with experience a few generalities emerge. In chromate-treated systems an inferior biocide may perform satisfactorily because of the leveling effect of the chromate ion itself on bacterial growth. Microbiological control is especially difficult in these circumstances: in non-chromate-treated systems, which are usually alkaline; in systems where the blowdown/volume ratio is less than .010, which in many systems corresponds to 6–8 cycles of concentration; in systems that have shown high plate counts (5–10 million bacteria per milliliter) for several weeks.

In Table 6.9 the effects of blowdown/volume ratio and weather on plate counts are shown for two cooling systems in the San Francisco Bay Area. Plate counts in the two columns head "rain" were measured during the month of January, when it rains almost daily. The counts in the two columns headed "dry, windy" were made during December, when the weather is cold, relative humidity is 35–40 percent, and prevailing winds are strong and from the north.

TABLE 6.9

Effect of Weather and Blowdown/Volume Ratio on Plate Counts

	Rain		*Dry, Windy*	
Days	No. 1 (b/V = .005)	No. 2 (b/V = .020)	No. 1 (b/V = .005)	No. 2 (b/V = .020)
0	2.0×10^6	20,000	2.7×10^6	2.0×10^6
2	*			*
6	220,000	20,000	6.5×10^6	47,000
8			**	
10	100,000	18,000	2.0×10^6	65,000
14	10,000	35,000	2.6×10^6	67,000
16			†	
20	100,000	55,000	1.3×10^6	260,000
26	130,000	50,000	1.7×10^6	1.0×10^6
31	180,000	100,000	300,000	2.5×10^6

* 40 ppm 10 percent methylene bisthiocyanate added day 2;
** 30 ppm added day 8;
† 40 ppm 35 percent dodecylguanidine hydrochloride added day 16.

The paucity of nutrients in cooling water limits bacterial counts in most instances to 2–3 million bacteria per milliliter, but it is advisable to keep plate counts below one million so that young bacterial cells predominate.

Young cells respire more rapidly, take in more toxicant, are more vulnerable, and succumb more readily than less rapidly respiring (or "resting") older cells. Resistances to certain biocides are developed by a small number of members of a bacterial population, but their number increases with time by a process of selecting out. A higher concentration or a different biocide is then required to control the new, more resistant strain.

Most bacteria that live in cooling water are relatively harmless, although sulfate-reducers can cause intensive corrosion. Chromate is toxic to these organisms, however, so they are found more often in systems on nonchromate treatments and in the anaerobic sediments at the bottoms of cooling tower basins, where they are inaccessible to biocides. Tehle,[31] in examining a number of different chromate-treated cooling waters, found sulfate-reducing bacteria in those taken from systems in which no biocides were used and the concentration of chromate was less than about 50 ppm. Some of these cooling waters contained high concentrations of sulfate (up to 4900 ppm) and most of those harboring sulfate-reducers were in the pH range of 6.7-7.2. The density of bacteria ranged from $10\text{-}1.6 \times 10^5$/ml in the waters, and from 900-200,000 in the basin sludges. Evidence of the presence of these organisms includes black particles, or the odor of hydrogen sulfide in the water, and pits containing a loose black deposit that rubs off easily leaving bright, shiny metal underneath. A special procedure is included in Chapter IX for detecting the presence of sulfate-reducing bacteria.

As remarked before, the most effective method for reducing bacteria count is a slug treatment, by which a lethal concentration of toxicant is added, all at once, to the recirculating water. Continuous treatment, which is sometimes advocated, is expensive and futile; this method should never be used. In many instances adding biocides continuously at 5-10 ppm only serves to engender a more troublesome, and more resistant population. Because of the continuous influx of dust, it is impossible to maintain a sterile open system, and it is extremely expensive to try.

Two of the more effective commercial biocides are methylene bisthiocyanate (7-10 percent) and dodecylguanidine hydrochloride (30-35 percent); the former is an enzyme poison, the latter disrupts cell walls and cytoplasm. The chemistry of both these agents has been discussed in Chapter V. Usually, methylene bisthiocyanate acts more rapidly than dodecylguanidine hydrochloride, and at a lower concentration of the active ingredient. Neither chemical, however, should be used at less than 30 ppm of the total formulation. Dodecylguanidine hydrochloride at 20 ppm total

(7 ppm active) actually stimulated the growth of bacteria in a cooling system. The initial plate count was 1.5×10^6 bacteria per milliliter, then after dodecylguanidine hydrochloride solution was added at 20 ppm, the plate count rose to 4.5×10^6. Part of this count, of course, is attributable to bacteria removed from surfaces by the scouring action of this cationic surfactant.

In Table 6.10 the influence of the concentration of methylene bisthiocyanate on viable plate counts is shown. This chemical also should not be used at less than 3 ppm active, nor at a pH greater than 7.5.

TABLE 6.10

Effective Concentration of Methylene Bisthiocyanate

Day	*Plate Count*	ppm MBT	*Day*	*Plate Count*	ppm MBT
0	2.3×10^6	–	21	–	21
1	–	10	24	1.0×10^6	–
6	1.8×10^6	–	26	–	32
8	–	21	27	200,000	–
9	1.0×10^6	–	33	25,000	–
13	700,000	–	37	150,000	–
19	1.8×10^6	–	40	360,000	–

Methylene bisthiocyanate and dodecylguanidine hydrochloride used alternately, not only keep microbiological growth under control, but also help keep a cooling system free from fouling. Dodecylguanidine hydrochloride is a powerful dispersant of inorganic salts and oxides, which once dispersed, remain suspended for many hours. Table 6.11 shows the results of using this chemical in a fouled system in which water had been treated for several years with a chromate-zinc-polyphosphate program with pH in the range of 7.0-7.5.

Upon examining the data for the first five-day period it is seen from the concentration ratios that the proportion of calcium in the recirculating water is increased with respect to magnesium, indicating that calcium scale is being dissolved. The increase in concentration of total phosphate shows that phosphate salts are also dissolving. The increase in plate count after treating the system with 35 ppm of dodecylguanidine hydrochloride solution, an amount normally toxic, shows that the biocide has been precipi-

TABLE 6.11

Effect of Dodecylguanidine Hydrochloride on a Fouled System

Day	*Recirculating Water* (ppm $CaCO_3$)			*Concentration Ratios**				
	TH	CaH	MgH	CaH	MgH	PO_4 (ppm)	*Dodecylguanidine* (ppm)	*Plate Count*
1	201	164	37	5.3	3.4	5	35	5.0×10^6
2	256	216	40	6.7	3.6	8	–	–
3	342	279	63	9.0	5.7	9	–	–
4	273	222	51	7.2	4.6	6	–	8.0×10^6
5	190	159	31	5.1	2.8	5	–	–
1	102	80	22	3.2	2.2	4	35	4.5×10^6
2	206	164	42	6.6	4.2	8	–	–
3	240	186	54	7.4	5.4	8	–	–
4	260	207	53	8.3	5.3	8	–	7.0×10^6
5	288	228	60	9.1	6.0	8	–	–
6	328	259	69	10.3	6.9	9	70	–
7	438	345	93	13.8	9.3	10	–	–
8	328	255	73	10.2	7.3	9	–	400,000
9	305	245	60	9.8	6.0	8	–	–

* Concentration ratio is the hardness in the recirculating water divided by that in the make up water. For the first five-day period the average hardness in the make up water was CaH, 13 and MgH, 11; during the second nine-day period CaH was 25 and MgH was 10.

tated, or otherwise rendered innocuous to bacteria.

After blowing down heavily for several days the blowdown valve was closed and dodecylguanidine was again added at 35 ppm, with results as before. When the dosage was doubled it is interesting to note that substantial amounts of both calcium and magnesium dissolved and that the plate count finally declined. As dodecylguanidine hydrochloride is cationic it forms insoluble salts with various inorganic anions when they are present at high concentration; normal levels of chromate and phosphate, however, do not precipitate the biocide.

When methylene bisthiocyanate and dodecylguanidine hydrochloride are used as described here, in conjunction with viable plate counts, microbiological control can be maintained without recourse to any other toxicants, chlorine included, although an equally effective program that is much less expensive is described in Section 5.5, below. If the pH of the water exceeds 7.5 chloroethylene bisthiocyanate should be substituted for methylene bisthiocyanate. In general, the chloro compound is more persistent, but it has an objectionable odor that sometimes limits its use.

Product data sheets for microbicides often specify treatment in ounces per thousand gallons, or in pounds per thousand gallons. These units can be converted to parts per million by multiplying by appropriate factors:

$$(\text{oz}/1000 \text{ gal}) \times 7.5 = \text{ppm}$$

$$(\text{lb}/1000 \text{ gal}) \times 120 = \text{ppm}$$

5.3 Quantitative Evaluation of Selected Microbicides

Thus far, the methods for controlling microbiological growth in cooling systems have been more-or-less empirical. The possibility of a quantitative measure of the efficiency of a microbicide can be considered extrapolated from a toxicant evaluation performed as described on page 200. This can be done satisfactorily with those enzyme poisons that are not appreciably affected by extraneous influences in the cooling system. The method to be described gives unreliable results with highly reactive toxicants such as chlorine, chlorine dioxide, and other chlorine derivatives, and to cationic biocides, including dodecylguanidine hydrochloride and quaternary ammonium compounds.

The three examples chosen are 10 percent methylene bisthiocyanate, 20 percent alkylaminopropane combined with 10 percent chlorophenol, and 20 percent 2,2-dibromo-3-nitrilopropionamide. The first two of these microbicides are persistent and neither is readily affected by miscellaneous debris, dirty screens, and the like. Therefore, their performance in a cooling system is more likely to approximate that in glassware.

The rate of disinfection of monocellular microorganisms is described by the following equation, which is sometimes known as Chick's law.[32]

$$\log_e(N_1/N_2) = k(t_2 - t_1) \tag{6.5}$$

where

N_1 is the number of microorganisms at time, t_1, when biocide is added.

N_2 is the number of microorganisms at time, t_2, some time after adding biocide.

k is the rate constant that depends upon the concentration of biocide.

Tables 6.12-6.14 list the number of microorganisms surviving exposure to various concentrations of the three selected microbicides for a period of six hours—the practical limit for maintaining a specific concentration of biocide in an operating recirculating cooling system. Values of the rate constant, k, are calculated from Eq. (6.5), then divided by the corresponding value of the concentration of toxicant to yield another constant, k'.

TABLE 6.12

Six-Hour Toxicant Evaluation of Methylene Bisthiocyanate

ppm Toxicant (c)	*Viable Plate Count*	*Survivors (%)*	*k*	*k'*
0	3,700,000	100	–	–
20	1,100,000	30	.202	.010
30	530,000	14	.324	.011
40	330,000	9	.403	.010
50	200,000	5	.486	.010

Equations can now be written to describe the action of the three biocides acting over a 6-h period.

For 10 percent methylene bisthiocyanate,

$$\log_e(N_1/N_2) = .010c(t_2 - t_1) \tag{6.6}$$

TABLE 6.13

Six-Hour Toxicant Evaluation of Alkylaminopropane Plus Chlorophenol

ppm Toxicant (c)	*Viable Plate Count*	*Survivors (%)*	*k*	*k'*
0	2,000,000	100	—	—
20	420,000	21	.260	.013
30	160,000	8	.421	.014
40	65,000	3	.571	.014
50	20,000	1	.768	.015

TABLE 6.14

Six-Hour Toxicant Evaluation of 2,2-Dibromo-3-nitrilopropionamide

ppm Toxicant (c)	*Viable Plate Count*	*Survivors (%)*	*k*	*k'*
0	1,400,000	100	—	—
2.5	740,000	53	.106	.042
5.0	360,000	26	.226	.045
7.5	170,000	12	.351	.047
10.0	90,000	6	.457	.046

For the alkylaminopropane-chlorophenol combination,

$$\log_e(N_1/N_2) = .014c(t_2 - t_1) \tag{6.7}$$

For 20 percent 2,2-dibromo-3-nitrilopropionamide.

$$\log_e(N_1/N_2) = .045c(t_2 - t_1) \tag{6.8}$$

If, for example, the plate count in a certain cooling system were found to be 6.0×10^6, the concentration of 10 percent methylene bisthiocyanate needed to reduce the plate count to, e.g., 1.0×10^5 in 6 h could be calculated:

$$\log_e (6.0 \times 10^6/1.0 \times 10^5) = .010c \times 6$$

$$4.094 = .060c$$

$$c = 4.094/.060$$

$$= 68 \text{ ppm}$$

This calculation takes into account that the higher the initial concentration of bacteria, the higher must be the concentration of biocide. Also, it is encouraging to note that the empirical values quoted in Table 6.10 are in reasonable agreement with values calculated from Eq. (6.6) for 10 percent methylene bisthiocyanate.

The above equations can also be used to estimate the number of survivors of a 6-h exposure to a certain concentration of biocide. Suppose there is a plate count of 6.0×10^6, and the advice given is to add 30 ppm of 10 percent methylene bisthiocyanate. Substituting the appropriate values in Eq. (6.6),

$$\log_e (6.0 \times 10^6) - \log_e N_2 = .01 \times 30 \times 6$$

$$15.6 - \log_e N_2 = 1.8$$

$$\log_e N_2 = 15.6 - 1.8$$

$$= 13.8$$

$$N_2 = 985{,}000$$

It would be best not to take this advice since 985,000 is an unacceptably high number of survivors. The correct dosage is 68 ppm, as calculated before.

If desired, the data in Tables 6.12-6.14 can be plotted as in Fig. 6.2 to show the dosage of each of the three biocides required to reduce viable plate counts from some initial value to 100,000 bacteria per milliliter within a 6-h period. Two-cycle, semi-logarithmic graph paper is suitable for the plots of plate counts against the concentration of biocide. Zero ppm of biocide, of course, corresponds to 0.1 million bacteria per milliliter, and because the curves are straight lines with slope $6k'$, only one other point need be calculated. As an example, suppose the plate count in some cooling system is found to be 5.0×10^6 bacteria per milliliter. Reduction of the count to 100,000 in 6 h is required, using 20 percent 2,2-dibromo-3-

nitrilopropionamide. Substituting the appropriate values in Eq. (6.8),

$$\log_e(5.0 \times 10^6/1.0 \times 10^5) = .045 \times c \times 6$$

$$3.912 = .27c$$

$$c = 3.912/.27$$

$$= 14.5 \text{ ppm}$$

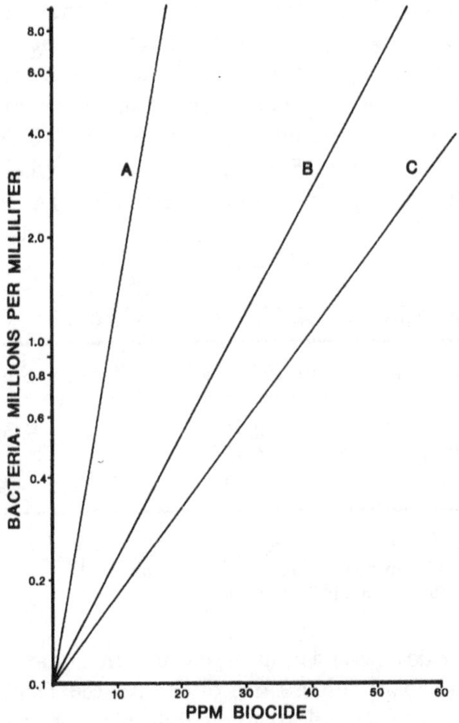

Fig. 6.2 Dosage of selected microbiocides; A = 20 percent 2,2-dibromo-3-nitrilopropionamide; B = 20 percent alkylaminopropane + 10 percent chlorophenol; C = 10 percent methylene bisthiocyanate.

This concentration is plotted against 5.0×10^6 in the second cycle of a sheet of semi-logarithmic graph paper and connected by a straight line to 1.0×10^5 at 0 ppm biocide. Similarly, the corresponding points for the other two biocides are calculated, obtaining the plots shown in Fig. 6.2.

Although dosages calculated from Chick's equation are, of course, not absolute values, those for methylene bisthiocyanate and the alkylamino-propane-chlorophenol combination work well in actual practice. Higher concentrations of 2,2-dibromo-3-nitrilopropionamide must sometimes be applied than those calculated from Eq. (6.8), probably because the compound is easily inactivated by contaminants.

The unit price of biocides, in itself, gives no indication of the cost of an effective treatment, but if it is assumed that the three biocides under discussion perform in a cooling system as they do in glassware, and if the unit prices of the three are known, it is possible to calculate the cost of treating a cooling system containing, say, 1.0×10^6 gal of water with each of the three biocides. If the bacteria count is found to be one million bacteria per milliliter, the costs of reducing this count to 1.0×10^5/ml using each biocide are shown in Table 6.15, calculated from Eqs. (6.6)-(6.8), or by using Fig. 6.2. The unit prices were those in effect in July, 1977.

TABLE 6.15

Relative Costs of Microbiological Control

Biocide	*Dosage* (ppm)	*Unit Price* ($/lb)	*Treating Cost* ($/10⁶ gal)
A	38.4	0.96	307
B	27.5	1.64	376
C	8.5	3.51	249

A = 10 percent methylene bisthiocyanate
B = 20 percent alkylaminopropane + 10 percent chlorophenol
C = 20 percent 2,2-dibromo-3-nitrilopropionamide

Again, while the dosages and costs in the table are not exact, experience has shown that in cooling systems free of debris, accumulations of slime, algal mats, oil, dirty screens, and so on, the three biocides perform approximately as indicated in the foregoing discussion.

Another way of comparing relative toxicities of microbicides is to determine their minimum inhibiting concentration when tested against individual species of microorganisms. This is not a practicable procedure for the average industrial laboratory; it is usually done in microbiological laboratories maintained by vendors or manufacturers of microbicides. Table 6.16 lists the minimum inhibiting concentrations of five microbicides against five microorganisms. *Pseudomonas aeruginosa* and *Aerobacter aerogenes*, as mentioned before, are common inhabitants of aqueous cooling systems, so the effect on these two species is of particular interest in screening biocides. *Bacillus myocoides* is a spore-forming bacterium not often of any significance in cooling systems. *Desulfovibrio desulfuricans*, an anaerobic sulfate-reducer, is of some concern, particularly in cooling systems on non-chromate treatments. *Aspergillus niger*, a rather resistant mold, is sometimes encountered in cooling systems, although whether it does any harm is a matter of discussion. For eradicating this mold, incidentally, 2,2-dibromo-3-nitrilopropionamide is superior to any of the biocides listed in Table 6.16.

5.4 Chlorination

Chlorine is favored as a biocide for large cooling systems, primarily because it is inexpensive. It does have disadvantages, however, the most serious being its deleterious effect on wooden towers, especially when the pH is over seven. Moreover, regulatory agencies are regarding chlorine in discharged water with increasing disfavor, so it probably will become necessary to dechlorinate, or else abandon chlorine entirely in streams discharged to public waters. In most places where chlorine is used it is fed into the suction of the recirculating pumps at a residual not exceeding 0.8 ppm in the warm return water. Comeaux and Tyler[33] recommend that chlorine be added directly to the basin at the point most remote from the pump suction, but not many do this as the gas tends to dissipate through the fan stacks causing odor complaints in populated areas. Control can be maintained by the usual o-tolidine method, or by measuring the oxidation potential of the recirculating water. Damage to tower wood is less severe if chlorine is used intermittently, but even so, occasional shock treatment with high concentration is likely to damage the wooden structure permanently.

When chlorine gas is bubbled into water hypochlorous acid, hydrogen ion, and chloride ion are formed:

$$Cl_2 + H_2O = HClO + H^+ + Cl^- \tag{6.10}$$

If the water contains weak bases, such as $CaCO_3$ or HCO_3^-, hydrogen ion reacts with them more readily than does the weak hypochlorous acid. Thus, the hydrolysis is favored by the presence of weak bases, which, in effect, make chlorine more soluble.

TABLE 6.16

Minimum Inhibiting Concentrations of Toxicants

		Toxicant (ppm)				
Toxicant	Active (%)	Pseudomonas aeruginosa	Aerobacter aerogenes	Bacillus mycoides	Desulfovibrio desulfuricans	Aspergillus niger
A	100	–	150	50	–	40
B	10	25	45	20	20	25
C	35	30	30	4	2	5
D	10	20	20	10	15	60
E	100	–	200	4	–	25

A = Acrolein
B = Chloroethylene bisthiocyanate
C = Dodecylguanidine hydrochloride
D = Methylene bisthiocyanate
E = Pentachlorophenol

Undissociated hypochlorous acid is the active toxicant in chlorination. Hypochlorite ion has no bactericidal properties, and it is for this reason that chlorine becomes rapidly ineffective as pH increases above 7. Ridge and Little[34] have investigated the equilibria in hypochlorite solutions and have calculated the fraction of undissociated HClO in the pH range from 1 to 10.

Cooling waters under pH control contain only small concentrations of weak bases, so the pH falls rapidly as chlorine is added. For instance, 2 ppm of chlorine added to a cooling system initially at pH 6.0, lowered the pH to 4.4. The effect varies, of course, with the buffering capacity of the recirculating water, the number of cycles of concentration, the cleanliness of the system, and other factors. In normal operating practice the acid injection is shut off and the pH is allowed to rise to 6.5-7.0 before chlorination is started. The rate of addition of the gas is then regulated so that the pH does not fall below 5.5.

As mentioned in Chapter V, chlorine is incompatible with many water-treating chemicals. It adds readily to a number of nitrogen-containing compounds, notably to ammonia, which is present as an impurity in some formulations of polyacrylate; to polyacrylamide; and to natural extractives in wood. Lignin is oxidized to a mixture of aldehydes and acids, ferrocyanide to ferricyanide, and 2-mercaptobenzothiazol is converted to the corresponding disulfide. In the presence of aminomethylenephosphonate and low pH, chlorine dissolves copper alloys; the cupric ion formed attacks iron.

Although Bringmann,[35] many years ago, showed by electron photomicrographs that chlorine causes no visible changes in the cell walls of bacteria, it is still alleged by some that chlorine "burns up" microorganisms. Chlorine, in fact, is ineffectual against masses of algae and slime, for it has little penetrating capability, although occasionally after chlorination there is a substantial increase in plate count. This seems to be caused by a loosening, or stirring up of accumulations that brings into solution many bacteria previously sessile. Dosage should be such that the "total residual chlorine," as measured by reaction with ortho-tolidine for three minutes, is not more than 2 ppm, and the "free available chlorine," as measured by reaction with the reagent for 5 s is 0.5-0.8 ppm. If the water treatment is compatible with chlorination the free available chlorine residual recommended should be attained within one to two hours, after which it should be maintained at 0.5-0.8 ppm for an additional hour.

To introduce chlorine into cooling water safely, a chlorinator is needed, the size of which depends upon the maximum recirculation rate of water through the system.

$$\text{Capacity (lb/24 h)} = \text{Recirculation rate (gpm)} \times \text{ppm dosage} \times 0.012 \tag{6.9}$$

Thus, for a cooling system recirculating at 50,000 gpm, the capacity needed to give 1-2 ppm of chlorine in the water is calculated by Eq. (6.9).

Chlorination rate, lb/24 h	*Chlorine Concentration* (ppm)
600	1.0
900	1.5
1200	2.0

From these values it can be seen that if chlorination proceeds for two hours and thirty minutes at the rate required to give 2.0 ppm total residual chlorine, the weight of chlorine added is 1200 × (2.5/24), or 125 lb. Chlorinating every other day, a one-ton cylinder lasts about a month. Excessive chlorine demand may indicate contamination of the cooling water by ammonia, monoethanolamine, or reducing agents.

5.5 Flocculant–Surfactant–Chlorine Dioxide Program

It is now possible to develop an economical program for microbiological control based on the principles discussed in the preceding sections of this chapter. In Section 1 of this chapter, it was noted that large numbers of bacteria adhere to surfaces within a recirculating system in the form of sticky films that tend to collect all manner of debris suspended in the water flowing over them. Sessile, adherent bacteria, of course, are not counted in viable plate counts made on samples of the recirculating water. Moreover, they are not accessible to most commercial microbicides until dislodged from the surfaces and dispersed as invisible flocs in the water. This can be done by adding a small amount of a cationic polyamine* having a molecular weight of 30,000-50,000, to the system each day.

Dixon and Zielyk[(36)] have investigated the flocculation of bacteria by polymers, measuring their effects on filtration rates, electrophoretic mobility, and light transmission and scattering. Polymers evaluated include polyacrylamide (nonionic) with molecular weights of 120,000-3,000,000,

* Magnifloc 521 CH, manufactured by the American Cyanamid Company, is suitable.

sodium polyacrylate (anionic) prepared by hydrolyzing the polyacrylamides, and polyethyleneimine (cationic) with molecular weights of 300 to 112,000. Nonionic polymers failed to flocculate *Escherichia coli* at pH values from 4-9, and concentrations ranging from 0.2-400 ppm of polymer; hydrolyzed polyacrylamide was also ineffective in similar experiments. The cationic polymers, however, were found to be efficient flocculants at concentrations from 0.1-50 ppm, although those of higher molecular weight caused bacteria to redisperse when used at concentrations in excess of 10 ppm. A polyethyleneimine with molecular weight of 21,000 was the most active flocculant when applied in the optimum concentration range of 1-10 ppm. Flocculation was inhibited to some extent at pH 4, and Dixon and Zielyk[(36)] point out that at this concentration of acidity a polymer must compete with hydrogen ions for adsorption sites on the surface of bacteria, and is unable to do so effectively if its molecular weight is too low. Also, with short polymers there is likely to be less bridging by the polymer from one bacterium to another. Conversely, with a long polymer chain multi-point adsorption is more likely, resulting in stronger attachment and more successful competition for sites on bacterial surfaces occupied by hydrogen ions. In the normal pH range of cooling water, say 6-8, the concentration of hydrogen ion is insufficient to exert a redispersing effect on bacteria so flocculation is unimpeded.

In Section 4 of this chapter the significance of the blowdown/total volume ratio, b/V, in microbiological control was demonstrated. Table 6.17 shows the effect of cationic polymer on viable plate counts as determined in system C, discussed on pages 121-125. In this cooling system b/V is equal to 0.0042, which corresponds to the extremely long half-depletion time of 165 h. Starting with day one, 1 ppm of cationic polyamine was added daily for a period of two weeks, during which time a large number of bacteria were dispersed into the recirculating water.

In Table 6.18 the opposite effect of the cationic polymer is shown. These results were obtained in system B (see pp 121-125), which has a b/V ratio of 0.014, thus permitting flocculated dispersed bacteria to be continuously flushed out of the system. In this instance no biocide was used for 55 days. Nevertheless, it is unwise to depend upon polymer entirely because a powerful cationic surfactant is required to dislodge those adherent bacterial films directly in contact with surfaces. Also, the polymer has negligible effect on algal mats within the cooling tower.

On the fifteenth day of the sequence in Table 6.17 6 ppm of dodecylguanidine hydrochloride was added to system C, followed on the sixteenth,

TABLE 6.17

Dispersion of Bacteria by Cationic Polymer

Day	*ppm Polymer added*	*Plate Count, bacterial/ml*
1	1	700,000
2	1	900,000
3	1	1,300,000
4	1	900,000
5	1	1,100,000
6	1	1,100,000
7	1	1,500,000
8	1	1,800,000
9	1	1,900,000
10	1	1,800,000
11	1	2,000,000
12	1	2,300,000
13	1	4,500,000
14	1	5,500,000

TABLE 6.18

Effect of *b/V* and Polymer on Plate Counts

Day	*ppm Polymer added*	*Plate Count, bacterial/ml*
1	1	260,000
2	1	750,000
3	1	540,000
4	1	800,000
5	1	560,000
6	1	360,000
7	1	230,000
8	1	200,000
9	1	160,000
10	1	250,000
11	1	160,000
12	1	140,000
13	1	130,000
14–55	1	< 100,000

seventeenth, and eighteenth days by 6-ppm portions of a 6 percent solution of stabilized chlorine dioxide. The plate count dropped to 400,000 per ml on the nineteenth day, then remained less than 100,000 per ml for about two weeks. For several days after this treatment the colonies that did develop were extremely small (almost pinpoint) indicating that a biostatic state had been achieved. This phenomenon is almost always observed after treatment with chlorine dioxide. The dioxide is much superior to chlorine on account of its persistence, and when used with a flocculant and a surfactant as described above, it is also superior to most nonoxidizing biocides; used alone it is not very effective. To recapitulate:

1. The system is treated with 1 ppm of cationic polyamine daily for from one to three days;
2. A 35 percent solution of dodecylguanidine hydrochloride is added at a total concentration of 6 ppm;
3. On each of the next three successive days 6 ppm of a 6 percent solution of chlorine dioxide is added, i.e., 0.36 ppm of chlorine dioxide.

This inexpensive program gives excellent results in cooling systems with b/V ratios ranging from 0.002-0.030, and it is highly recommended.

At this point it is worthwhile to calculate the total concentration of the polymer being added to system B at the rate of 1 ppm every 24 h. This can be done by iterative solution of the depletion equation,

$$\log_e (c/c_0) = -bt/V \tag{6.11}$$

where

$$b/V = 0.014$$
$$t = 24 \text{ h}$$

For example,

$$\begin{aligned} \log_e c &= \log_e c_0 - 0.014 \times 24 \\ &= \log_e c_0 - 0.336 \\ &= \log_e (1) - 0.336 \\ &= 0 - 0.336 \\ c &= 0.7146 \end{aligned}$$

On the second day 1 ppm of polymer is added, c_0 becomes $1.00 + 0.71$, or 1.71 ppm, and so on. As c_0 increases daily the amount of polymer lost in the blowdown also increases until the amount lost each day is equal to the amount added, i.e., 1 ppm. The computed values of c_{24} and c_0 are shown in Table 6.19.

TABLE 6.19

Concentration of Polymer in System B

Day	c_0 (ppm)	c_{24}, *Residual after 24 h* (ppm)	*24-h loss* (ppm)
1	1.00	0.71	0.29
2	1.71	1.22	0.49
3	2.22	1.59	0.63
4	2.59	1.85	0.74
5	2.85	2.04	0.81
6	3.04	2.17	0.87
7	3.17	2.26	0.91
8	3.26	2.33	0.93
9	3.33	2.38	0.95
10	3.38	2.41	0.97
11	3.41	2.44	0.97
12	3.44	2.46	0.98
13	3.46	2.47	0.99
14	3.47	2.48	0.99
15	3.48	2.49	0.99
16	3.49	2.49	1.00

It would be desirable to be able to calculate the value of c when the steady state concentration is reached, and also the number of days required to reach the steady state, without the labor of an iterative solution of Eq. (6.11). To do this it is noted that at the steady state the concentration of polymer ranges in 24 h between a maximum concentration, c_s, and a minimum concentration, c_m, with a difference between them of 1 ppm. That is,

$$c_s = c_m + 1$$

$$c_s - 1 = c_m = 0.7146 c_s$$

$$c_s - 0.7146c_s = 1$$

$$(1 - 0.7146)c_s = 1$$

$$c_s = 1/(1 - 0.7146)$$

$$= 3.50 \text{ ppm}$$

Table 6.19 indicates that the maximum concentration of polymer under the conditions given is 3.49 ppm, but this is because the values in the table are rounded off to only two decimal places in each day's calculation.

The reader familiar with geometric progressions[37] will recognize the term $1/(1 - 0.7146)$ as being of the form $a/(1 - r)$, which is the limiting value upon which the sum of the infinite geometric progression,

$$a + ar + ar^2 + \ldots + \infty$$

converges, when r is less than 1. That is,

$$S = \lim_{n \to \infty} \frac{a(1 - r^n)}{(1 - r)} = \frac{a}{(1 - r)} \tag{6.12}$$

In our example $r = e^{-.336}$. If Δ equal the daily increment in concentration of polymer, i.e., 1 ppm, the concentration of polymer on each successive day can be expressed in terms of Δ and r.

Day	c_t
1	$0 + \Delta = \Delta(1)$
2	$r\Delta + \Delta = \Delta(1 + r)$
3	$\Delta + r(r\Delta + \Delta) = r^2\Delta + r\Delta + \Delta$
	$= \Delta(1 + r + r^2)$

and so on. Thus, a geometric progression can be formulated:

$$a + ar + ar^2 + \ldots + \infty = a/(1 - r) \tag{6.13}$$

with $a = 1$. Now if the progression consists of a finite number of terms, the sum of the series is

$$a + ar + ar^2 + \ldots + ar^n = \frac{a(r^n - 1)}{(r - 1)} \tag{6.14}$$

Also, when $r < 1$

$$\frac{a(r^n - 1)}{(r - 1)} = \frac{a(1 - r^n)}{(1 - r)} \tag{6.15}$$

In this example, then, the steady state concentration of polymer is

$$c_s = \frac{\Delta(1 - r^n)}{(1 - r)} \tag{6.16}$$

where n is the number of days from the start of the treatment. Eq. (6.16) is easily solved for n by rearranging and taking logarithms.

$$n = \log_e \left[1 - \frac{c_s(1 - r)}{\Delta}\right] / \log_e r \tag{6.17}$$

Since it was seen in Table 6.19 that the limiting concentration appears to be 3.49 when the terms are carried to only two decimal places, this value will be used for c_s.

$$n = \log_e \left[1 - \frac{3.49(1 - 0.7146)}{1}\right] / \log_e (0.7146)$$

$$= \log_e [0.0039540/\log_e (0.7146)]$$

$$= (-5.533)/(-0.3360)$$

$$= 16.5 \text{ days}$$

If the terms in Table 6.19 are carried to four decimal places the concentration of 3.5000 is reached in 20.27 days. For practical purposes the difference between 3.49 and 3.50 is insignificant. It is intriguing that the concentration of a chemical being added in regular increments while at the same time undergoing logarithmic depletion is explicitly defined by a geometric progression.

6. EFFECTIVE APPLICATION OF RECOMMENDED MICROBICIDES

The results of a number of years of experience in the application of microbicides to cooling water can now be summarized. Throughout the book examples are given of the effects of various biocides used in particular ways. In several instances what was done was not necessarily what should have been done, in light of subsequent experience. Therefore, it is appropriate now to provide the reader with directions for using those biocides that have been found most economical and effective over a long period of time. The recommended chemicals, all of which are available either from the various water service companies or direct from the manufacturer, are:

1. A stabilized solution containing 6 percent of chlorine dioxide, used with a cationic polymer and a surfactant, as described in Section 5.5;
2. A 35 percent solution of dodecylguanidine hydrochloride;
3. A mixed formulation containing approximately 20 percent of alkylaminopropane and 10 percent of 2,4,5-trichlorophenol;
4. A 10 percent solution of methylene bisthiocyanate;
5. A 20 percent solution of 2,2-dibromo-3-nitrilopropionamide.

For day-to-day routine microbiological control the five-day successive treatment with polymer, surfactant, and stabilized chlorine dioxide, as described in Section 5, pp. 144–147, is effective and economical. The optimum concentration of a 6 percent solution of chlorine dioxide is 6 ppm, i.e., 0.36 ppm of the active ingredient. This solution does not work well used alone, nor is it effective in the presence of oil or reducing agents such as hydrogen sulfide and sulfur dioxide.

A 35 percent solution of dodecylguanidine hydrochloride is used as a surfactant at 6 ppm in the chlorine dioxide program. When applied alone as a general microbicide, however, the proper concentration range is 30–40 ppm. *Pseudomonas* and *Aerobacter* are destroyed by 20 ppm of the 35 percent solution, but as the chemical is cationic, it attaches to negative sites by electrostatic bonding, rapidly decreasing its concentration in cooling water. For this reason, an excess of 50–100 percent must be used to allow for depletion. Dodecylguanidine hydrochloride at 2 ppm is toxic to *Desulfovibrio desulfuricans*, and is the microbicide of choice should this organism be encountered. Green and blue-green algae are killed by 2-5 ppm of the chemical, while at the same time the structure of the cooling tower is thoroughly cleaned by the scouring action of the surfactant. By attaching itself to carboxylate groups in cellulose, the compound protects wetted parts of the cooling tower from invasion by fungi. The chemical is rapidly inactivated, and thus inefficient, in fouled or dirty systems.

A mixed formulation of alkylaminopropane and trichlorophenol functions well in the presence of organic material, dirt, mud, and microbiological debris, and also in the presence of oil, provided the system is first treated with 5 ppm (active) of sodium dioctylsulfosuccinate. The recommended concentration range in dirty systems is 25–45 ppm. Because of the persistence of chlorophenols this formulation should be used with discretion.

Methylene bisthiocyanate at a concentration of 10 percent is a generally applicable, nonacclimatizing biocide that is useful as a follow-up treatment when others fail to lower plate counts sufficiently. The proper concentration range in a clean system is 40–60 ppm; the chemical is not useful in the presence of oil.

TABLE 6.20

Cost Comparison of Treatments with Recommended Biocides

Biocide	*Effective Dosage* (ppm)	*Pounds required*	$/lb	*Cost per treatment* ($)
Dodecylguanidine hydrochloride (35 percent)	40	141.8	1.85	262.26
Alkylaminopropane (20 percent) plus 2,4,5-trichlorophenol (10 percent	36	127.6	1.64	209.27
Methylene bisthiocyanate (10 percent)	50	177.2	0.96	170.14
2,2-dibromo-3-nitrilopropionamide (20 percent)	11	39.0	3.51	136.85
Cationic Polymer (Day 1)	1	3.5	0.52	1.84
Dodecylguanidine (Day 2)	6	21.3	1.85	39.40
Chlorine Dioxide (Day 3)	6	21.3	0.94	20.00
Chlorine Dioxide (Day 4)	6	21.3	0.94	20.00
Chlorine Dioxide (Day 5)	6	21.3	0.94	20.00
				101.24

A relatively new biocide is 2,2-dibromo-3-nitrilopropionamide. At

10-15 ppm this compound is toxic to certain highly resistant molds that sometimes infest recirculating cooling systems and are unaffected by the other toxicants.

Table 6.20 shows dosages and costs of treating a 425,000-gal cooling system, containing 2.0×10^6 bacteria per milliliter, with the chemicals discussed in this section. Unit prices depend upon the quantity purchased and the price structure of the supplier; the values shown in the table are those that were in effect for single-barrel purchases in July of 1977. As with most commodities, the lowest prices for biocides are obtained by competitive bidding.

Microbiological control of contaminated systems requires special procedures that are discussed separately in Chapter VII.

References

(1) Potter, L. F. 1964. "Planktonic and benthic bacteria in lakes and ponds." In *Principles and applications in aquatic microbiology*, chapter 8, eds. H. Heukelekian and N. Dondero, New York: Wiley.

(2) Meadows, P. S. and Anderson, J. G. 1966. Microorganisms attached to marine and freshwater sand grains. *Nature* 212: 1059.

(3) Stotzky, G. 1973. Techniques to study interactions between microorganisms and clay minerals *in vivo* and *in vitrio. Bull. Ecol. Res. Comm.* (Stockholm) 17:17.

(4) Zo Bell, C. E. 1943. The effect of solid surfaces upon bacterial activity. *J. Bacteriol.* 46:39.

(5) Bott, T. L. and Brock, T. D. 1970. Growth and metabolism of periphytic bacteria methodology. *Limnol. Oceanogr.* 15: 333.

(6) Sharpley, J. M. 1966. *Elementary petroleum microbiology.* Gulf Publishing Company, Houston, Texas.

(7) Bungay, H. R., III. 1968. Microbial interactions in continuous culture. In *Advances in applied microbiology*, vol. 10, eds. W. W. Umbreit and D. Perlman, New York and London: Academic Press.

(8) Edwards, B. C. May, 1969. Role of polymeric flocculants in sediment removal in condenser cooling systems. *J. Appl. Chem.* 19:141.

(9) Tenney, M. W. and Stumm, W. 1965. Chemical flocculation of microorganisms in biological waste treatment. *J. Water Pollution Control Federation* 37:1370.

(10) Busch, P. L. and Stumm, W. Jan. 1968. Chemical interactions in the aggregation of bacteria. *Environmental Science and Technology* 2:1, 49.

(11) American Public Health Assoc. 1976. *Standard methods for the examination of water and wastewater.* 14th ed., New York.

(12) Winogradsky, S. 1923. *Sur la méthode directe dans l'étude microbiologique du sol. C. R. Acad. Sc.*, Paris, 177:1001.

(13) Conn, H. J. Culture media for use in the plate method of counting soil bacteria. Tech. Bull. N.Y. St. Agric. Exp. Sta. No. 38.

(14) Egdell, J. W., Cuthbert, W. A., Scarlett, C. A., Thomas, S. B., and Westmacott, M. H. 1960. Some studies of the colony count technique for soil bacteria. *J. Appl. Bact.* 23:1, 69.

(15) American Petroleum Institute. May, 1959. RP 38, *Recommended practice for biological analysis of water-flood injection waters.* Publication of the Division of Petroleum.
(16) Chambers, C. W. and Clarke, N. A. 1966. Control of bacteria in nondomestic water supplies. *Advan. Appl. Microbiology* 8:105.
(17) Cotton, R. A., Sladek, K. J., and Sohn, B. I. 1975. Evaluation of a single-step bacterial pollution monitor. *J. Am. Water Works Assoc.* 67:449.
(18) Collins, C. H. 1964. *Microbiological methods.* London: Butterworths.
(19) Holmsen, H., Holmsen, I., and Bernhardsen, A. 1966. Microdetermination of adenosine diphosphate and adenosine triphosphate in plasma with the firefly luceiferase system. *Anal. Bio-Chem.* 17:456.
(20) Stanley, P. E. and Wilkins, S. G. 1969. Use of the liquid scintillation spectrometer for determining ATP by luciferase enzyme. *Anal. Biochem.* 20:381.
(21) Van Dyke, U., Stitzel, R., Clellton, T., and Szustkiewica. 1969. An automated procedure for the sensitive and specific determination of ATP. *Clinical Chemistry* 15:No. 1, 3.
(22) Reid, R. and Leese, H. Feb. 6, 1975. A new form of biological power. *New Scientist* 65:310.
(23) Potter, L. F. 1960. The effect of pH on the development of bacteria in water stored in glass containers. *Can. J. Microbiol.* 6:257.
(24) Salle, A. J. 1973. *Fundamental principles of bacteriology.* 7th ed., New York: McGraw-Hill.
(25) Foster, J. W. 1949. *Chemical activities of fungi.* New York: Academic Press.
(26) Brock, T. D. March, 1971. Microbial growth rates in nature. *Bact. Rev.* 35:1, 39.
(27) McCoy, J. W. 1974. *The chemical treatment of cooling water.* New York: Chemical Publishing Company.
(28) Postgate, J. 1969. *Microbes and man.* Baltimore: Penguin Books.
(29) McCoy, J. W. 1962. *The inorganic analysis of petroleum.* pp. 198–200, New York: Chemical Publishing Company.
(30) Yost, W. H. Apr. 16, 1973. Microbiological control in recirculating water systems avoids fouling. *Oil and Gas J.* 71:16, 107.
(31) Tehle, E., Jr. 1966. Sulfate reducing bacteria in water cooling systems. *Mater. Prot.* 5:12, 21.

(32) Chick, H. 1908. An investigation of the laws of disinfection. *J. Hyg.* (Cambridge) 8:92.

(33) Comeaux, R. V. and Tyler, C. P. Oct., 1962. *Oxidation reduction potential control of cooling tower chlorination.* Proceedings of the 23rd International Water Conference.

(34) Ridge, B. P. and Little, A. H. 1942. The control of pH in hypochlorite solutions. Parts I and II. *J. Textile Institute* 33: T33.

(35) Bringman, G. 1953. Electron microscopic findings of the action of chlorine, bromine, iodine, copper, silver and hydrogen peroxide on *Escherichia coli. Z. Hyg. Infektionskrankh.* 138:155.

(36) Dixon, J. K. and Zielyk, M. W. 1969. Control of bacteria content of water with synthetic polymer flocculants. *Environmental Science and Technology* 3:551.

(37) Weast, R. C. and Selby, S. M. eds. 1970. *Handbook of tables for mathematics.* 4th ed., Cleveland: Chemical Rubber Company.

VII.

Microorganisms in Contaminated Water

Mishaps, misoperations, and equipment failures occur now and then in the chemical, food, and petroleum processing industries. As these often lead to the contamination of cooling water by various chemicals, the effects of some of these on the microorganisms that inhabit cooling systems are now considered, and appropriate responses to potentially harmful leaks are suggested.

1. INFLUENCE OF CONTAMINANTS ON MICROBIOLOGICAL GROWTH

Life is difficult for microorganisms in recirculating cooling water, for under normal operating conditions they are continuously on the verge of starvation, their environment undergoes radical changes of temperature every few minutes, and they are periodically menaced by toxic chemicals added for the purpose of destroying them. Repeated temperature shocks have an adverse effect on cell division, and disturb the synthesis of deoxyribonucleic acid. Thus, Sherman and Cameron[(1)] found that 95 percent of young cells of *Escherichia coli* are killed by rapid cooling from 45 to 10 C, although gradual cooling over a period of 30 min is harmless. For other reasons this particular organism cannot survive long in cooling water, but *Pseudomonas aeruginosa*, a common species in soil and cooling water, shows a similar sensitivity to cold shock.

Unlike higher plants and animals, bacteria quickly adapt to extreme changes in their surroundings, and with very little encouragement can continue to proliferate. They readily adjust their enzyme systems to take advantage of the food available, tenaciously coping with their environment as they find it. For instance, the rate of respiration of soil bacteria is increased markedly by many easily assimilable chemicals such as acetaldehyde and other aldehydes, methanol, ethanol, and glycol. This is accompanied by an increase in the rate of reproduction, so if any of these chemicals are introduced into cooling water, bacteria counts will rise.

In the majority of uncontaminated open recirculating cooling systems the concentration of bacteria in the water is naturally limited to about 2.0×10^6 organisms per milliliter, as measured by viable plate count. Harrison,[2] in studying the growth response of *Bacterium lactis aerogenes* in the absence of nutrient, found that cells in the logarithmic phase of growth (see page 37) showed very high death rates at a concentration of about 1.0×10^6 cells per milliliter, but at the critical concentration of 1.0×10^8 cells per milliliter the death rate eventually became negligible. This he attributed to a loss of cellular substances (phosphate, ammonia, nucleic acids, etc.) which at low cell densities are too dilute to be reassimilated. The same phenomenon probably places a natural limit on the population of uncontaminated cooling water.

2. *COMMON CONTAMINANTS AND THEIR EFFECTS*

Contrary to general opinion, many contaminants do not noticeably affect the living organisms in cooling water. Liquid and gaseous paraffins, lubricating oils, waxes, hydrogen sulfide, and oxides of nitrogen leaking into a cooling system have little immediate influence on the growth of microorganisms. Edible oils, such as cottonseed, coconut, flaxseed, sesame seed, and peanut oil foul a cooling system badly, but it takes time for the microbial population to shift to species that can metabolize these complex mixtures of glycerides. The most troublesome growths are those engendered by alkanolamines, alcohols, glycols, urea, and aromatic hydrocarbons. These chemicals support and encourage a number of harmful species not normally encountered in cooling water; these are discussed in detail later in this chapter.

2.1 Dirt

Although dirt must be considered a contaminant in open recirculating cooling systems, its ubiquitousness ensures that it will inevitably be present in all systems, together with its resident microorganisms. Most of the organisms in soil reside in the surface layers, which are swept up by the wind and drawn into cooling towers. The number and variety of organisms vary with the seasons, the moisture content of the soil, the proportion of humus, and other factors. Algae, fungi, and bacteria are always on hand in great numbers, however, as well as protozoa, worms, rotifers, and other higher forms. The soil bacteria that grow on agar plates are distributed approximately as follows: actinomycetes (filamentous bacteria) 10-40 percent; nonspore-forming bacteria 50-80 percent; spore-forming bacteria 3-10 percent. Cooling water is not a satisfactory medium for many of these species, however, and they are quickly crowded out by the hardier *Aerobacter* and *Pseudomonas*; the latter can metabolize at least 75 different organic chemicals.

Throughout the book the presence of soil bacteria has been assumed, and the biocidal treatments already described cope with those organisms that are in or in contact with the recirculating water. A substantial amount of mud eventually accumulates on the bottom of the cooling tower basin, where in time it becomes a breeding ground for microorganisms that are inaccessible to biocides in the water. Protozoa, rotifers, and ciliates abound in basin sludge, and anaerobic bacteria grow within and beneath it, generating in extreme cases objectionable odors and unmanageable masses of slime. This nuisance can only be abolished by mechanical cleaning. Once a year sludge should be removed from the basin by a vacuum truck; this can be done by a cooling tower contractor while the tower is in operation, if the deposit of sludge is not too thick.

2.2 Hydrocarbons

As mentioned earlier, paraffins are only slightly soluble in water, and therefore are not easily accessible to microorganisms. Furthermore, only a few species of *Mycobacterium*, *Penicillium*, and *Pseudomonas* manufacture paraffin oxidase, an enzyme that is essential for metabolizing paraffins. If leaks persist, however, microbes that are able to metabolize the leaking chemical increase in number, and may become a nuisance.

It is of interest to note that recombinant biology[3] has permitted the introduction into a single bacterium of specific genetic traits from four different strains of bacteria that naturally metabolize certain hydrocarbon fractions. The new bacterium is a strain of the *Pseudomonas* genus that can metabolize some two-thirds of the components of crude oil at several times the rate of other oil-consuming bacteria, some of which may take over a year to consume oil. The new strain was created by inserting plasmids from each of the four strains into a single strain. Plasmids are double-stranded rings of deoxyribonucleic acid, found outside the chromosome, that carry the genetic code of the digestive enzymes and confer resistance to antibiotics, among other functions. Their combination in a single bacterium produces an individual cell that is nutritionally more versatile than its progenitors.

Oil contaminants can be identified by extracting them from cooling water with Freon 113,* trichlorotrifluoroethane, then subjecting the extract to chromatographic analysis, comparing the chromatogram obtained with those of various streams in the petroleum processing plant, or plants, under suspicion. A standard method[4] has been published for making this identification.

Because of the tendency of bacteria to concentrate at oil-water interfaces, bacteria counts in the presence of oil may not reflect the actual number of bacteria in the system. Occasionally, very stable oil-water emulsions are formed by the homogenizing action of the recirculating pumps. This is especially common with light and medium cycles oils produced in catalytic cracking and used as fuel oil cutters. To the unaided eye these emulsions appear to be homogeneous slimes, but under the microscope minute oil droplets can be seen. These are bright under low focusing and dark under high focusing because their refractive index is less than that of water. In other words, the oil droplets appear to be optically empty. Microbiological activity is, in general, more intense in these intimate mixtures of oil and water than it is beneath oil films.

The mononuclear aromatic hydrocarbons benzene, cumene, ethylbenzene, toluene, and the xylenes, cause severe fouling for they are carbon sources for filamentous bacteria such as *Sphaerotilis natans*, whose growth is facilitated by the appreciable solubility in water of the aromatic hydrocarbons. Filamentous bacteria form tough, rubbery-textured, malodorous masses in the presence of xylenes that tend to collect on the surface of the water in the basin. The sticky masses plug screens tightly, and they are exceedingly difficult to wash off. Once established, these organisms cannot be controlled until the aromatics leak is stopped. Further details are given in Section 3 of this chapter.

2.3 Edible Oils

Margarine factories and plants that process vegetable oils are frequently plagued with heavy deposits of fats and oils in the basins of cooling towers and within their cooling systems. As these accumulations are tenacious and impervious to emulsifiers, microorganisms develop that can metabolize fats and oils. Fats are converted to glycerol and fatty acids by enzymes called esterases. These products then provide nourishment for other microorganisms that are unable to use the complex compounds themselves, so

* Manufactured by E. I. DuPont de Nemours & Company, Inc., Wilmington, Delaware.

that an enormous increase in the number of bacteria ensues. Sexsmith, et al.,[5] suggest the use of unspecified polyelectrolytes to disperse fats and oils.

2.4 Nitrogen Compounds

Ammonia is a common contaminant in recirculating cooling systems associated with chemical processing and petroleum refining. Several soil bacteria including *Proteus vulgaris*, *Pseudomonas aeruginosa*, and *Aerobacter aerogenes* are able to synthesize their amino acids from ammonia, provided suitable sources of carbon and other growth factors such as biotin, phosphate, and oxygen are also available. As there is normally a paucity of nutrients in cooling water, the presence of 25-200 ppm of ammonia has no significant effect on the bacterial density as measured by viable plate count. In a system continuously contaminated by 25-50 ppm of ammonia, however, there will usually be found a concentration of 2-10 ppm of nitrite, which is formed in the process of nitrosofication carried out by members of the autotrophic genera *Nitrosomonas*, *Nitrosocystis*, *Nitrosogloea*, *Nitrosospira*, or *Nitrosococcus*. These microorganisms obtain their energy by oxidizing ammonia to nitrite in neutral, or very slightly alkaline medium, using carbon dioxide or carbonates as their sole source of carbon.

$$OH^- + NH_3 + 3/2\,O_2 = NO_2^- + 2H_2O \qquad (7.1)$$

Other bacteria, *Nitrobacter* and *Nitrocystis*, oxidize nitrite to nitrate.

$$NO_2^- + 1/2O_2 = NO_3^- \qquad (7.2)$$

All of the nitrifying bacteria coexist in soil and are the principal source of the nitrate therein. *Nitrocystis* and *Nitrosogloea* can be particularly troublesome in cooling systems as they both form zoogloea–that is, masses of cells embedded in slime.

Plate count agar does not support the growth of autotrophic bacteria that oxidize ammonia to nitrite, so their number is not determined by the viable plate count method described in Chapter VI. A special procedure for preparing an autotrophic medium is given in Chapter IX that is useful for maintaining microbiological control of cooling systems associated with the manufacture of ammonia.

Approximately 2 percent of the microorganisms in soil can hydrolyze urea; common species are *Aerobacter ureae*, *Micrococcus ureae*, and *Sar-*

cina ureae.

$$O = C(NH_2)_2 + H_2O = 2NH_3 + CO_2 \qquad (7.3)$$

These species are likely to be found in cooling water in urea manufacturing or processing plants. In common with the nitrifying bacteria, they do not grow on plate count agar because, of course, there is no urea in that medium.

2.5 Alkanolamines

Both carbon dioxide and hydrogen sulfide are acid gases that participate in a reversible acid-base reaction with monoethanolamine and diethanolamine. These chemicals, therefore, are used for purifying nitrogen, removing hydrogen sulfide from propane, removing carbon dioxide from hydrogen produced by steam reforming of methane, and for sweetening refinery fuel gas from thermal cracking units. Monoethanolamine is preferred to diethanolamine for sweetening gas because, of the two, it is the stronger base, has the lower equivalent weight, and is more readily purified by distillation.

$$HOCH_2CH_2NH_2 + CO_2 + H_2O = HOCH_2CH_2NH_3^+HCO_3^- \qquad (7.4)$$

$$HOCH_2CH_2NH_2 + H_2S = HOCH_2CH_2NH_3^+HS^- \qquad (7.5)$$

These reactions are reversed when the temperature is raised to 250-300 F.

As gas processing plants are usually constructed of carbon steel, corrosion by hydrogen sulfide or carbon dioxide is often seen on the hot ends of heat exchanger tubes. Stress corrosion cracking also is fairly common. In addition, corrosion rates are increased by accumulations of sludge, high concentrations of monoethanolamine, and inefficient operation of the reclaimer, so the perforation of tubes often occurs, allowing amine to leak into the cooling water. Prolonged leaks of monoethanolamine encourage the proliferation of spore-forming bacteria, which are otherwise rare in recirculating cooling systems. If these leaks are ignored for several days, a resistant, flocculant mold may appear in the water that is extremely difficult to eradicate. The serious consequences of alkanolamine leaks are discussed further in Section 3 of this chapter.

2.6 Hydrogen Sulfide

When hydrogen sulfide leaks into cooling water it is rapidly oxidized to sulfur by oxygen, or by chromate, if that chemical is being used as a corrosion inhbitor. Free sulfur is visible as a milky colloidal solution in the recirculating water. Should the leak continue for several days, sulfur bacteria such a *Thiobacterium* may appear. These are very large rod-shaped cells containing granules of elemental sulfur; they oxidize hydrogen sulfide, sulfur, thiosulfate, or sulfite to sulfate. As sulfur enters into the composition of all living cells, and is thus essential to all organisms, it is likely that other microorganisms are stimulated by colloidal sulfur in cooling water. At any rate viable plate counts usually increase several-fold after a leak of hydrogen sulfide, even though sulfur bacteria cannot grow on agar plates.

A small leak of hydrogen sulfide can be handled for a short time by adding chlorine.

$$H_2S + 4Cl_2 + 4H_2O = 10H^+ + SO_4^{--} + 8Cl^- \qquad (7.6)$$

Note, however, that ten moles of hydrogen ion are formed per mole of hydrogen sulfide. For this reason, only very small leaks can be controlled with chlorine, otherwise the pH of the recirculating water will fall too low.

2.7 Glycols

Water can be removed from natural gas by taking advantage of its tendency to form hydrogen bonded intermolecular associations with glycols. Water must be removed from gas to prevent the formation of snowy, crystalline hydrates of methane, ethane, propane, and iso-butane. These hydrates, which resemble water crystals and form at high pressures in wet, turbulent locations, cause difficulties in the pumping and distribution of natural gas. Mixtures of glycols are also used in petroleum refining–for example, in the UDEX process for separating aromatic hydrocarbons.

In the presence of glycols, the lipid content of the cell walls of bacteria increases and they become more impervious to biocides. Also, these chemicals enhance the slime-forming propensities of some bacteria. There is always a rapid rise in the concentration of the resident bacteria when cooling water becomes contaminated with glycols.

3. *SPECIFIC TREATMENTS FOR CONTAMINATED SYSTEMS*

A great many difficulties can be avoided by maintaining a careful surveillance of cooling systems, watching especially for early indications of contamination of the water by process chemicals, or by gases from the atmosphere. Foreign odors in the recirculating water or near the cooling tower deserve special attention. Several chemicals that can be identified by their odors are ammonia, hydrogen sulfide, sulfur dioxide, phenols, alcohols, light hydrocarbons, benzene, and xylenes.

Iridescence on the surface of a sample of cooling water may suggest a leaking lubricating oil cooler. Most petroleum products are insoluble in water, and thus tend to collect in a layer on the surface of the water in the basin of the cooling tower. Small leaks of oil give the water a milky appearance, as does contamination by hydrogen sulfide. Minor leaks of hydrogen sulfide that occur within a cooling system may not impart any odor to the water in the basin because the gas is rapidly oxidized by oxygen, or by chromate in a chromate-treated system. Some odor is usually detectable, however, around the fan stacks on top of the cooling tower.

Chromate-treated cooling water turns green if sulfur dioxide is drawn into the cooling tower. This gas reduces the yellow chromate ion to the green chromic ion. Hydrogen sulfide, on the other hand, produces a grayish-green precipitate of hydrous chromic oxide combined with colloidal sulfur. The disappearance of the yellow color of chromate ion, or an abnormally high requirement of the inhibitor to maintain the desired residual, always indicates the presence of some reducing agent—typically sulfur dioxide, hydrogen sulfide, or some organic chemicals—or else a low pH from an overdose of sulfuric acid. Changes in pH in either direction may also signal the presence of a contaminant. Abnormally high pH is caused by leaks of bases such as ammonia, primary amines, monoethanolamine, or diethanolamine. Sulfur dioxide, a leaking chlorinator, and in some circumstances, hydrogen sulfide, produce low pH values.

Small leaks are insidious, for although not obvious, they are often the precursors of severe microbiological fouling. Examples are continuous leaks of a few parts per million of xylenes, benzene, monoethanolamine, or diethanolamine. Failure of a routine biocide program to reduce viable plate counts should always prompt an immediate search for small leaks of some process chemical into the cooling water.

In the previous chapter heavy emphasis was laid on viable plate counts as an essential feature in any rational program for microbiological control, this being the only way to tell when biocide is needed and whether it has done any good after it has been used. We must now temper our position by pointing out that there are many soil bacteria that cannot grow on plate count agar. Among these are species that fix nitrogen, oxidize sulfur, reduce sulfate, hydrolyze urea, oxidize ammonia, or digest cellulose. Plate count agar contains glucose, tryptone, and yeast extract. It is an enriched medium that supports the growth of zymogenous organisms, that is, those that cause fermentation, but it inhibits many other organisms indigenous to soil. The monosaccharide glucose is a readily available source of energy and carbon, especially agreeable to the fermenters. Tryptone, a peptone formed by the action of the proteolytic enzyme pancreatin on the milk protein casein, serves as a source of nitrogen for many bacteria. Yeast extract is a rich source of the B vitamins thiamine, riboflavin, pyridoxin, and pantothenic acid, and as such is a potent stimulant of bacterial growth. Autotrophic bacteria, however, must have ammonia or ammonium salts, for they are unable to use peptones as a source of nitrogen. Similarly, bacteria that hydrolyze urea can only grow in the presence of urea. Thus, ammonia or urea leaking into cooling water can give rise to enormous numbers of bacteria that do not develop any colonies on agar plates. Furthermore, they crowd out bacteria that do grow on plate count agar, suggesting to the unwary that all is well with respect to microbiological control. This is an important consideration in operating cooling systems associated with fertilizer and ammonia manufacturing plants.

Certain contaminants (e.g., sugars and glycols) cause the concentration of lipids to rise in the cell walls of bacteria, strengthening their resistance to toxicants. As these lipids are firmly bound to protein and carbohydrate molecules, they are not directly extractable with fat solvents. Oily and fatty contaminants, acting as protective colloids, film bacteria and shield them from toxicants as well as from adverse environmental effects including heat, cold, and low or high pH. Surfactants are available, however, that emulsify oil and increase the permeability of cell walls. Thus, it is often advantageous to formulate microbicides with surfactants (also called surface tension depressants or wetting agents), or alternatively, to use one of these chemicals as a pretreatment, or adjunct, to enhance the susceptibility of protected microorganisms to a toxicant. Also, bacteria enclosed in fatty or waxy capsules seem to be more vulnerable to biocides at alkaline pH values.

Normal microorganisms, too, are damaged to some extent by surfactants. Microbial cells are complex aqueous colloidal systems within which numerous boundaries exist between colloidally dispersed intracellular compounds and the aqueous solution. Surface tension depressants–phenol and Aerosol OT, for example–accumulate at these interfaces, decreasing the forces that maintain the boundaries and increasing the permeability of the cytoplasmic membrane so that amino acids and other vital material diffuse out of the cell. Surfactants that raise the interfacial tension within a living cell cause colloidal particles to coagulate, thereby inactivating the organism's enzymes.

As mentioned in Chapter V, the destruction of bacteria and other single-celled organisms by chemicals is logarithmic, that is, the number of organisms killed in a given interval of time is greatest when the toxicant is first added, then becomes steadily less as disinfection continues. For this reason, a microbicide of some persistence is required to gain control of the growth of microorganisms in a contaminated cooling system. Enzyme poisons, for instance, are sometimes ineffective because the synthesis of every enzyme is regulated by a specific nucleoprotein. Therefore, if the controlling gene is unharmed, its corresponding enzyme is replaced as soon as the toxicant disappears. This is why acrolein is not very useful as a microbicide–it is so reactive that it rapidly disappears from the scene. To be lethal a toxicant must inactivate a critical enzyme system for a relatively long time, i.e., several hours.

Quick response is necessary to contaminants that are likely to promote biological growth. Young cells, in the logarithmic growth phase (see page 37), are much more susceptible to microbicides than are senescent organisms, which respire slowly and are likely to be enclosed in protective mucilagenous capsules. It is, therefore, advisable to make an early kill should cooling water become contaminated by readily metabolized chemicals, especially aromatic hydrocarbons, alkanolamines, or glycols. Several remedial treatments that can be used to discourage microbiological growth in contaminated cooling water are discussed next.

3.1 Petroleum Products

Lubricating oil for pumps, compressors, and other machines often finds its way into cooling water through leaks in oil coolers. Recirculating cooling systems serving petroleum processing plants are continually exposed to a variety of different petroleum products including naphthas, gas oils, cycle oils, and residua. Naphthenes and polynuclear aromatics are not readily degraded by soil bacteria, but normal paraffins are oxidized successively to primary alcohol, aldehyde, and aliphatic acid.[6] *Pseudomonas*, for instance, oxidizes ethane to ethanol, acetaldehyde, and acetic acid; a small amount of carbon dioxide is also released. Chromatograms of oils extracted from cooling water invariably show a deficiency in the concentration of paraffins compared to that in the original petroleum fraction, manifested by a pronounced attenuation of the paraffin elution peaks. Biological activity during oil contamination is seldom intense, however, as other necessary growth factors are usually missing or deficient.

Other things being equal, the rate of oxidation of paraffins is proportional to their solubilities in water. Table 7.1, based on data published by McAuliffe,[7] shows the solubilities at room temperature of several different hydrocarbon compounds. As an example, opposite carbon number six in the table the solubilities are shown for n-hexane (10 ppm), 1-hexene (50 ppm), 1,5-hexadiene (170 ppm), and 1-hexyne (360 ppm). For a given carbon number, solubility in water is increased by ring formation, by increasing unsaturation, and in paraffins and olefins, by branching.

TABLE 7.1

Solubilities of Hydrocarbons in Water

	Solubility, ppm			
Carbon Number	n-Paraffin	1-Olefin	Diolefin	Alkyne
5	39	150	560	1570
6	10	50	170	360
7	3	15	44	94
8	0.7	3	10*	24

* Estimated value.

With few exceptions, petroleum leaks that are detected and repaired within a day or two have no perceptible effect on the microbiological population in cooling water. After a leak has been stopped, however, it is important to remove all oil from the cooling system, as it coats fill slats, sticks to the walls of the basin, fouls screens, and because of its affinity for dirt, causes fouling within heat exchangers.

The most efficient method for cleaning up after oil leaks is to add to the cooling water 5 ppm of the oil emulsifier sodium dioctylsulfosuccinate. This chemical is sold in an isopropanolic solution under the trade name

Aerosol OT,* as well as other rebrands. Let us suppose, for illustration, that a light cycle oil has leaked into the cooling system of a fluid catalytic cracking unit. The total volume of cooling water is 425,000 gal. The following remedial procedure is recommended.

1. After the leak has been found and stopped, have any oil on the surface of the water in the basin skimmed and drawn into a vacuum truck.
2. Close the blowdown valve, and add 10 ppm of a 50 percent solution of sodium dioctylsulfosuccinate. (The number of pounds to add, in this case, is $10 \times 425{,}000 \times 8.33/10^6 = 35$ lb. The solution weighs about 11 lb/gal, so 3.2 gal are required.)
3. Allow to circulate for 4 h (there may be copious foam) then open the blowdown valve and flush overnight.

Light cycle oil is virtually insoluble in water, but as it contains 50-60 percent of heavy aromatics in the carbon range of 11-16, the number of bacteria in the water often begins to increase three or four days after a leak of this material. It is, therefore, advisable after flushing overnight to add to the water about 40 ppm of a 10 percent solution of methylene bisthiocyanate, or alternatively, 6 ppm of a 6 percent solution of chlorine dioxide on each of three successive days.

If the source of the leak cannot be found immediately it is better to add emulsifier daily while maintaining a heavy blowdown than to pour in expensive biocides. As already mentioned, oil coats microorganisms in the water, so chlorine has little effect, whether added continuously or intermittently. Until the leak is repaired blowdown is the most effective and economical method for limiting the microbial population.

3.2 Mononuclear Aromatic Hydrocarbons

Cooling systems associated with aromatics recovery plants, catalytic reformers, xylene isomerization plants, and paraxylene manufacturing are subject to contamination by benzene, toluene, ethylbenzene, and the xylenes. These mononuclear aromatic hydrocarbons are particularly troublesome as they are appreciably soluble in water, where they support the growth of filamentous bacteria. Table 7.2 records the sequence of events resulting from a small amount of ortho-xylene continuously leaking into

* Manufactured by American Cyanamid Company, Process Chemicals Dept., Wayne, New Jersey.

cooling water in a catalytic reformer plant. It is a daily record of bacteria counts, microbicides used, and concentrations of ortho-xylene in the cooling water during March, April, and May in the year 1975.

Toward the end of February bacteria counts suddenly rose above one million per milliliter. The routine treatment with cationic polyamine, dodecylguanidine hydrochloride, and chlorine dioxide described on page 144, was carried out from March 3rd to 7th without reducing the concentration of bacteria in the water. At the same time, a sticky, malodorous mass of filamentous bacteria, resembling *Sphaerotilis natans*, began to form on the surface of the water in the basin. On March 10th, 56 ppm of a proprietary formulation of alkylaminopropane and chlorophenol were added, which temporarily effected a significant reduction in the bacteria count. By March 15th, however, the count was back up to 4.0×10^6 per milliliter. Two other microbicides, 10 percent methylene bisthiocyanate and 35 percent dodecylguanidine hydrochloride were added successively, which held the count to about 1.0×10^6 per milliliter.

During the last week in March it was belatedly realized that the bacterial growth was being supported by the small concentration of ortho-xylene dissolved in the cooling water. Operating commitments precluded shutting down immediately for repair, so a holding action was elected whereby 6 ppm of a 6 percent solution of chlorine dioxide was added to the cooling water daily. Through the 16th of April this treatment succeeded in holding the bacteria counts to about one million, or a little less. The floating tangle of filamentous bacteria remained on the water, but its mass did not sensibly increase.

On April 18th the leak was repaired, but on April 26th the count was again over one million. The surfactant-chlorine dioxide routine was again carried out, with the discouraging result of raising the count to 1.7×10^6 per milliliter, although while this treatment was going on the floating masses of bacteria disintegrated and disappeared. Methylene bisthiocyanate was once again used, and this time the system was finally brought under control. Xylene, through the month of May, was not detected in the water; in the ensuing months an occasional routine treatment was sufficient to maintain control.

Whenever repeated routine treatments with microbicide fail to lower bacteria counts the existence of a leak of some easily metabolized material should be suspected and measures taken to have it repaired as soon as practicable. The foregoing example illustrates the futility of pouring expensive biocides into cooling water contaminated with an easily metabolized

TABLE 7.2

Sequence of Events During a Xylene Leak

March	*Bacteria/ml*	*Biocide* (ppm)	*Xylene* (ppm)	*April*	*Bacteria/ml*	*Biocide* (ppm)	*Xylene* (ppm)	*May*	*Bacteria/ml*	*Biocide* (ppm)	*Xylene* (ppm)
1	4,000,000			1	1,200,000	CD 6		1		CD 6	
2			17	2	1,000,000	CD 6		2		CD 6	
3	4,200,000	CPA 1		3		CD 6		3	1,700,000	CD 6	
4	4,900,000	DGH 6		4		CD 6	15	4			Nil
5	2,800,000	CD 6		5	1,100,000	CD 6		5		MBT 40	
6	1,700,000	CD 6		6	1,200,000	CD 6		6			
7	3,100,000	CD 6		7	650,000	CD 6		7			
8	4,000,000			8	1,300,000	CD 6		8			
9			16	9	1,200,000	CD 6		9			
10	4,000,000	AAP 56		10		CD 6		10	580,000		
11	600,000			11		CD 6	14	11			
12	240,000		14	12	930,000	CD 6		12	570,000		
13				13	710,000	CD 6		13			
14				14	690,000	CD 6		14			
15	4,000,000		11	15	800,000	CD 6		15			

TABLE 7.2 *(cont'd.)*

March	*Bacteria/ml*	*Biocide* (ppm)	*Xylene* (ppm)	*April*	*Bacteria/ml*	*Biocide* (ppm)	*Xylene* (ppm)	*May*	*Bacteria/ml*	*Biocide* (ppm)	*Xylene* (ppm)
16	4,500,000			16	750,000	CD 6		16			
17	2,800,000	MBT 40	11	17				17	340,000		
18				18	(xylene leak repaired)			18			
19		DGH40		19	600,000		5	19			
20	130,000			20	600,000			20			
21	1,000,000			21				21			
22	1,200,000		14	22	720,000			22			
23	1,000,000			23	680,000			23			
24		AAP 40		24				24	460,000		
25	600,000	CD 6		25				25			Nil
26	500,000	CD 6		26	1,200,000		2	26			
27	280,000	CD 6		27				27			
28	900,000	CD 6		28				28			
29	1,000,000	CD 6		29		CPA 1		29	100,000		
30	1,300,000	CD 6	18	30		DGH 6		30			
31		CD 6						31			

AAP = 20 percent Alkylaminopropane + 10 percent chlorophenol; CD = 6 percent Chlorine dioxide; CPA = Cationic polyamine; DGH = 35 percent Dodecylguanidine hydrochloride; MBT = 10 percent Methylene bisthiocyanate.

chemical. The total cost of biocides during this fiasco, incidentally, was \$1710 in March, \$557 in April, and \$200 in May.

The series of plate counts from March 6th through March 8th, when the counts increased daily despite the addition of chlorine dioxide, deserves comment. This phenomenon is attributed to variations in operating conditions in the processing plant served by the cooling system. Table 7.3 lists some values that reflect these operations.

TABLE 7.3

Cooling Water Conditions

Date	*Make up* (gph)	*Cooling Range* (°F)	*Chromate* (ppm)	*Cycles*
3/6/75	26,617	27	15	5.6
3/7/75	31,246	33	15	6.6
3/8/75	39,892	38	15	8.4

The rise in make up rate and cooling range show that during this period the output of the plant, and thus the evaporation rate, were being steadily increased. On each of these three days 16.2 lb of chromate were added in one portion to maintain a concentration in the range of 15-20 ppm of chromate. As the total volume of the system is 425,000 gal, 16.2 lb of chromate is equivalent to $(16.2 \times 10^6)/(425{,}000 \times 8.33)$, or 4.6 ppm. Thus, the concentration of chromate over a 24-h period ranged from 15 to (15 + 4.6), or 19.6 ppm. From this, using the depletion equation, the blowdown rate can be calculated, which obviously was constant during the three-day period under discussion.

$$\begin{aligned} \log_e(15/19.6) &= -b \times 24/425{,}000 \\ -0.2675 &= -b \times 24/425{,}000 \\ b &= 0.2675 \times 425{,}000/24 \\ &= 4737 \text{ gph} \end{aligned}$$

The values of cycles of concentration tabulated in Table 7.3 were then calculated, noting that cycles = make-up rate/blowdown rate. Thus, the cycles of concentration and the temperature of the water were simultaneously increasing, both of which favor the multiplication of bacteria.

3.3 Alkanolamines

Monoethanolamine or diethanolamine leaking into cooling water invariably cause reactions that curtail heat exchange. Both chemicals are highly alkaline and so precipitate calcium, magnesium, and iron. In addition, both compounds reduce chromate to a gelatinous slime of hydrous chromic oxide. If the leak is large, or prolonged, a flocculent, resistant mold develops that demands repeated heavy dosages of biocide to bring it under control. The mold, which resembles *Aspergillus*, consists of a mycelial matrix with rounded sporophores, or fruiting bodies, containing fungal spores. The outer integument of the spores is an amorphous, chitin-like polysaccharide that is impervious to most chemicals. As these molds are large forms, they form visible flocs in the cooling water.

Table 7.4 contains the details of a massive contamination of water in a 300,000-gal cooling system by monoethanolamine. On day one the system was operating normally with a negligible concentration of bacteria in the cooling water. On day seven the pH of the cooling water suddenly began to rise so rapidly that the acid pump used to control pH was unable to keep up. A sample of recirculating water was checked and it was determined that it contained 1500 ppm of monoethanolamine–about 3750 lb in this system.

The addition of chromate was discontinued immediately and not begun again until the leaking cooler had been removed from service and the system had been thoroughly flushed. In the event, there was a two-day delay before the leaking heat exchanger was blocked-in.

Next, dodecylguanidine hydrochloride was added at a concentration of 40 ppm. This was done with the objective of minimizing fouling by hydrous chromic oxide and the oxidation products of monoethanolamine and also to discourage the growth of bacteria. The cooling system was blown down heavily for three or four days, during which time the cooling water appeared normal. On the fourteenth day, however, the water acquired a noticeable cloud, so a supplemental charge of dodecylguanidine hydrochloride was added on this and the next two successive days. On the fifteenth day the concentration of bacteria was 840,000/ml, with 50 ppm of monoethanolamine still remaining in the recirculating water. The cloud in the water continued to deepen, then on the seventeenth day flocs were observed and it became necessary to clean the screens ahead of the recirculating pumps every 8 hours.

TABLE 7.4

Sequence of Events During a Monoethanolamine Leak

Day	*Bacteria/ml*	*Biocide, ppm*	*MEA, ppm*	*Condition of Water*
1	100,000			Clear
7		DGH 40	1500	
8	300,000			
14		DGH 20		Noticeably cloudy
15	840,000	DGH 20	50	Noticeably cloudy
16		DGH 20		Heavy cloud
17		AAP 40		Floc in water–screens plugging
18	2,000,000	AAP 40		Floc in water–screens plugging
21			36	Dispersed flocs
22		AAP 40		Dispersed flocs
23	1,300,000		10	Clearing
29		CPA 1	Nil	Clear
30		DGH 6		
31		CD 6		
32		CD 6		
33		CD 6		
36	200,000			

AAP = 20 percent Alkylaminopropane + 10 percent Chlorophenol; CD = 6 percent Chlorine dioxide; CPA = Cationic polyamine; DGH = 35 percent Dodecylguanidine hydrochloride; MEA = Monoethanolamine.

The situation by this time was more-or-less out of hand. Accordingly, a proprietary mixture of alkylaminopropane and chlorophenol was added on three separate occasions while the system was under continuous blowdown. On the twenty-third day the water finally began to clear. The regular routine program with surfactants and chlorine dioxide was instituted on the twenty-ninth day, and when it was completed, both the appearance of the water and the viable plate count were normal. The total cost of biocides used as the result of this mishap was $1028.

Note that it took ten days after the contamination for the plate count to rise above one million, although the mold began to grow about one week after the mishap. This demonstrates again that bacteria do not immediately begin to flourish after every instance of contamination of cooling water by organic chemicals. Indeed, this entire sequence of events was caused by delay in reducing the concentration of monoethanolamine in the cooling water, further evidence of the need for rapid response to mishaps. Obviously, the use of heavy blowdown here played a significant role in regaining control of microbiological growth, but a further complication was the greatly augmented biochemical oxygen demand (BOD) in the oxidation pond that received the blowdown water. Had 2,2-dibromo-3-nitrilopropionamide been available, it would undoubtedly have been more effective in eradicating the mold than the sequence actually used here.

3.4 Hydrogen Sulfide

Hydrogen sulfide is a by-product of many petroleum processes, notably fluid catalytic cracking, catalytic reforming, crude oil distillation, isocracking, hydrogenation, and hydrocracking. When this gas dissolves in cooling water treated with a chromate-zinc corrosion inhibitor, gelatinous sludges of hydrous chromic oxide, zinc sulfide, and elemental sulfur precipitate immediately, and rapidly foul heat exchangers, especially those in which the leaks occur. In general, if the leak is allowed to continue, microbiological growth begins to accelerate after 3–4 days of contamination by colloidal sulfur.

Upon discovering a leak of hydrogen sulfide, the addition of chromate should be discontinued until the leak has been repaired. The initial reaction with chromate is represented by the following equations.

$$2CrO_4^{--} + 3H_2S + 2H_2O = 2Cr(OH)_3 + 3S + 4OH^- \qquad (7.7)$$

Hydrogen sulfide also precipitates zinc as the sulfide. Thus, it destroys both anodic and cathodic corrosion protection. Furthermore, it reacts directly with iron to form ferrous sulfide and hydrogen.

$$H_2S + Zn^{++} = ZnS + 2H^+ \qquad (7.8)$$

$$H_2S + Fe = FeS + H_2 \qquad (7.9)$$

From the three preceding equations it is apparent that hydrogen sulfide leaking into cooling water is intensely corrosive. For this reason, every effort is made to isolate and repair leaks of hydrogen sulfide without delay, and if sulfur is cleared from the system by blowdown, there are seldom any difficulties with microbiological growth. If plate counts do increase, one or two 40-ppm doses of either dodecylguanidine hydrochloride or methylene bisthiocyanate are usually sufficient to restore the system to normal.

3.5 Ammonia

Occasional contamination of cooling water by ammonia at concentrations less than 100 ppm has no significant effect on microbiological growth. In cooling systems serving ammonia manufacturing and fertilizer plants, however, where ammonia is continuously present in the water, autotrophic bacteria develop and sometimes become a nuisance.

Special mineral media are required to support the growth of nitrifying bacteria. For these bacteria, 25 C is the preferred incubation temperature and silica gel is superior to agar as the solidifying agent.(8) Silica gel plates of autotrophic medium are troublesome to prepare, and as these bacteria are usually confined to the special plants mentioned, these counts are seldom made. Directions are provided in Chapter IX for those interested in making counts of nitrifying bacteria. In general, the proliferation of nitrifiers is limited by the lack of other essential factors such as phosphate and the trace metals boron, cobalt, copper, and manganese. Thus, regular viable plate counts can be used for routine microbiological control.

Because of the high chlorine demand imposed by the presence of ammonia, chlorine is ineffective for dealing with microorganisms in cooling systems associated with ammonia manufacturing plants. Chlorine dioxide is well suited to these systems, however, as is methylene bisthiocyanate.

References

(1) Sherman, J. M. and Cameron, G. M. 1934. Lethal environmental factors within the natural range of growth. *J. Bacteriol.* 27: 341.
(2) Harrison, A. P., Jr. 1960. The response of *Bacterium lactis aerogenes* when held at growth temperature in the absence of nutriment: an analysis of survival curves. *Proc. Roy. Soc., Ser. B* 152:418.
(3) Anonymous, Sept. 22, 1975. New microbe designed for oil spill cleanup. *Chem. and Eng. News*, 53, 8.
(4) Am. Soc. Testing Materials. 1976 Book of ASTM Standards. *Comparison of waterborne petroleum oils by gas chromatography*, Part 31, D3328-74aT: 585.
(5) Sexsmith, D. R., Silverstein, R. M., and Savinelli, E. A. Oct. 27–29, 1970. *Polymers in cooling water treatment.* Proceedings of the 31st Annual Water Conference of the Engineer's Society of Western Pennsylvania: 141–150.
(6) Davis, J. B. 1967. *Petroleum microbiology.* Amsterdam: Elsevier.
(7) McAuliffe, C. 1966. Solubility in water of paraffin, cycloparaffin, olefin, cycloolefin, and aromatic hydrocarbons. *J. Phys. Chem.* 70:1267.
(8) Funk, H. B. and Krulwich, T. A. 1964. Preparation of clear silica gels that can be streaked. *J. Bacteriol.* 88:1200.

VIII.

Optical Microscopy

Practitioners in the field of cooling water treatment are prone to describe every slimy, gelatinous, or amorphous residue from a cooling system as microbiological. Calling a substance microbiological, however, does not make it so. To identify such fouling positively it is necessary to resort to the microscope, by which means most slimes can be unequivocally characterized. An elaborate instrument is not essential, nor is any special training required beyond the manufacturer's instruction manual. A great deal can be seen with a simple magnifier capable of 100 X magnification, but of course, a compound microscope with 10 X, 20 X, and 40 X objectives has much greater range and flexibility. Also, bacteria can be seen more clearly at the higher magnifications, although it is seldom necessary to use a 100 X oil-immersion objective.

There is no purpose in trying to identify specific genera of bacteria, algae, and other microorganisms, as the only interest now is to assess the effectiveness of microbicides in cooling systems. The intent of this chapter is not to train the reader to be a microscopist, but rather to show the uninitiated how to distinguish microbiological fouling from the many other types. It is futile, and very expensive, to try to remove accumulations of clay gels, for example, by massive doses of microbicides under the impression that the fouling is microbiological. The detailed examination of fouling material is covered later in Section 2 of this chapter.

1. METHODS

A preliminary examination at low magnification should always be made, even when using a compound microscope, to get some idea of the gross appearance of particulate matter. At high magnification it is impossible to view large particles totally in focus.

With the exception of amoeba and some fungi, practically all microorganisms have regular shapes, with one or two axes of symmetry. That is, they are rod-shaped, filamentous, circular, or of some other geometric form, as distinguished from particles of soil, hydrous oxides, silt, and inorganic chemical slimes, all of which are either irregularly shaped or amorphous. Oil droplets, which are common in cooling water, appear under the microscope as bright disks, each surrounded by a dark circle.

Brief directions are given here for preparing wet slides, staining bacteria, and examining various types of fouling. For detailed discussions of the compound microscope and its accessories the reader is referred to a work by Chamot and Mason.[1] Special bacteriological techniques, such as the staining of spores, capsules, and flagella are described in a handbook by Baker and Breach.[2]

1.1 Wet Slides

Algae are best observed in aqueous medium under low magnification (100 X). To confirm the presence of algae, transfer a small amount of the material to be examined to a 25 X 75-mm glass microscope slide by means of an inoculating loop. Add a drop or two of distilled water, cover with a 25-mm No. 2 glass cover slip, and place the slide on the microscope stage. Using a 10 X ocular and a 10 X objective lens focus on the material on the slide. Most algae from cooling towers will appear as a tangle of segmented filaments, usually with entrapped particles of dirt and often with adhering droplets of oil. Samples from within a cooling system rarely contain viable filamentous algae, although fragments of skeletons of filaments are common in deposits from stagnant areas. Color plates in the Standard Methods[3] illustrate some of the algae most frequently encountered.

1.2 Staining Bacteria

Unless a phase contrast microscope is available it is necessary to stain bacteria before they can be clearly seen under bright-field illumination. It is fairly simple, and highly instructive, to prepare a stained smear of bacteria from a colony grown on an agar plate, as described in Chapter IX. To do this, a minute portion of a bacterial colony is transferred to a clean glass slide. A small amount of water containing a nonionic wetting agent is added to the tiny smear on the slide and the mixture is spread uniformly over the working area of the slide. The bacteria are then fixed by gentle heating over a light flame. The slide is next drenched with Gram crystal violet stain, then the excess stain is rinsed off with cold water, and the stained bacteria are examined under a microscope using a 10 X ocular and a 20 X or 40 X objective, if available.

The purpose of this exercise is to permit the reader to see for himself the enormous number of bacteria that make up a colony, and more importantly, to show how to distinguish microbiological fouling from other accumulations. Hughes[4] has estimated the average weight of a bacterium as 10^{-12} g; that is, 10^9 bacteria weigh about 1 mg. If 1 mg of bacteria are spread over the working area of a slide ($25 \times 50 = 1250$ mm^2) there will be $10^9/1250$, or 800,000 bacteria/mm^2. The area of the field observed through an objective lens depends on its magnifying power and on the diameter of the diaphragm in the eyepiece in the plane of the real image,[1] but for a 40 X objective the diameter of the field is about 0.35 mm and its area is about 0.10 mm^2. Therefore, approximately 80,000 tiny rods are visible when a slide prepared by the following procedure is viewed through a 10 X ocular and a 40 X objective.

PREPARATION

1. *Equipment:* a) Microscope slides, 25 X 75 mm, with one end frosted on one side; working area 25 X 50 mm; Gold Seal No. 3050, Becton, Dickinson and Company, Parsippany, New Jersey 07054.
   b) Nichrome inoculating loop, 3-mm loop.
2. *Reagents:* a) Gram crystal violet stain, Difco No. 3329-76-7, Difco Laboratories, Detroit, Michigan 48232.
   b) Stock solution of wetting agent. Transfer 1.0 g of Triton X-100 (Rohm & Haas Company, Philadelphia, Pennsylvania 19105) to a 1000-ml volumetric flask, and dilute to volume with water.
   c) Dilute solution of wetting agent (5 ppm). Transfer 5.0 ml of stock solution to a 1000-ml volumetric flask and dilute to volume with water.

PROCEDURE

1. Touch the surface of a bacterial colony on an agar plate very lightly with the edge of an inoculating loop, then transfer the minute smear to the center of the working area of a clean glass slide.
2. Add two drops of the dilute solution of wetting agent to the slide and mix the smear and solution thoroughly, spreading the mixture uniformly over the slide.

3. Allow the water to evaporate, then flame to fix the bacteria on the slide. (The slide should never be more than just warm to the touch.)
4. Drench the slide with Gram crystal violet stain, then let stand for one minute.
5. Rinse immediately with cold water, dry the slide, and examine it under the microscope.

Soil bacteria that produce slime surround themselves with a capsule of polysaccharide molecules that is four to five times the volume of the living cell; in general, the capsule is an integral part of the organism that remains attached to the cell wall. Capsules can be made visible by special staining techniques. In one method alcian blue is used, which in basic solution forms a blue complex with hydroxyl groups in polysaccharide molecules. The dye molecule is too large to penetrate the cytoplasmic membrane of living cells, but when a cell dies its permeability increases so that internal polysaccharides are also stained. This sometimes confuses the interpretation of results. Also, young rapidly respiring cells have such thin capsules that they cannot be seen. In practice carbolfuchsin is used as a counterstain so that capsular material is stained blue and the protoplasm within the cell is stained red. The interested reader may refer to other sources[2,5] for detailed procedures.

2. *EXAMINATION AND IDENTIFICATION OF FOULING MATERIAL*

As mentioned before, there is a tendency to describe any gelatinous deposit or floc in cooling water as "bio." It is true that filamentous growths of bacteria, notably *Sphaerotilis natans*, appear on the surfaces of rivers and streams, either free-floating or attached to various surfaces, and that in time these growths disintegrate and become detached, after which small flocs can be seen floating on the water for some time afterward.[6] When suspended floc in cooling water is chemically analyzed, however, it is usually found to be composed of inorganic material, often zinc phosphate, chromic hydroxide, or clay. Similarly, scums and foams on cooling water more often than not are caused by chemicals.

Although turbidity is frequently attributed to the presence of biological material, an extremely high density of bacteria is required to produce a visible haze in water—1.0×10^7 or more per milliliter, in fact. Five hundred million cells per milliliter form a cloudy, but not highly turbid suspension.

If gelatinous sludges are examined under the microscope it is easy to tell whether they are composed of hydrated dirt, inorganic oxide gels, oil-water emulsions, bacterial slimes, or algae, although these are normally undistinguishable by the unaided eye. Especially misleading is the appearance of various forms of hydrated soil. The ideal soil, if there is such a thing, consists of 35 percent sand, 25 percent clay, 10 percent humus, and 30 percent water.[7] The organic portion of a soil includes sugars, cellulose, hemicellulose, pectin, fats, waxes, and proteins derived from the decomposition of vegetation, all of which is collectively called humus.

Strictly speaking, however, humus is the end product of the decay of felled timber, stumps, forest litter, leaf mold, animal manures, roots, weeds, fruits and vegetables, and a great many other precursors. The final product is a colloidal mixture that increases the water-holding capacity of clay, silt, sand, and the like.

Oglesby, et al.[8] have discussed many aspects of the slow transformation of lignins derived from higher plants to humus through the biological action of *Basidiomycetes*, the fungi that are responsible for white-rot in cooling towers. They estimate that in six to eight months 60–75 percent of the lignin in natural substrates is transformed to humus on the forest floor. Humus is not a specific organic compound, but rather a mixture of humin, fulvic acid, hymatomelanic acid, and humic acid; the latter usually predominates and may be a metabolic end product excreted by fungi. According to Alexander,[9] other products of lignolysis by fungi are coniferaldehyde, p-hydroxycinnamaldehyde, ferulic acid, 4-hydroxy-3-methoxyphenylpyruvic acid, p-hydroxycinnamic acid, guaiacyl glycerol, vanillin, vanillic acid, and dehydrodivanillin.

Radiocarbon dating[9] has shown some samples of humus to be nearly 3000 years old. The material is thus remarkably resistant to biological oxidation, a property attributable to the presence of an aromatic polymer. A large proportion of organically bound phosphorus in humus, in the form of phosphates of polysaccharides and inositol, is available with difficulty to microorganisms; the nitrogenous components are also extremely resistant to biological oxidation. It might be expected that polysaccharides would be easily metabolized by microorganisms, yet 10–15 percent of stable organic matter in soils is polysaccharide.

Humus is responsible for the color of soil, and in addition to its capacity to retain water; it is also the agent that holds and releases nutrients and carbon dioxide to the surrounding soil. Dry plant residues in soil can absorb up to 300 percent of moisture,[10] forming gelatinous masses in quiescent parts of cooling systems. Clays are so finely divided that they readily form colloidal solutions, or sols, in cooling water that under certain conditions undergo gelation, setting to a more-or-less homogeneous, firm, elastic gel.[11] These gels contain 2-3 percent clay with the remainder water. Particles of clay also tend to adhere to each other when wet and deposit as impervious sticky masses. Oil leaks further complicate matters, as clay has a marked affinity for oil. Weiser[12] has found that when highly dispersed and precipitated slowly, particles of hydrous chromic oxide form a fairly rigid gel that contains only 0.18 percent Cr_2O_3. These gels are common in chromate-treated cooling systems subjected to small leaks of reducing agents—in particular hydrogen sulfide. As these gels are green, they are often erroneously supposed to be algal growths.

All of the foregoing gelatinous materials, formed from particles that absorb water strongly, have been described on occasion as microbiological, although the most cursory examination under magnification would prove that judgment incorrect. Microbiological growths, in general, possess three characteristics that can be used to identify them: they are stringy; they are sticky; and they smell bad. Bacterial slime invariably has a characteristic odor similar to that of bacterial colonies on an agar plate. Also, under magnification, bacterial slime looks exactly the same as a smear from a bacterial colony; i.e., myriads of small straight rods, most of which are alive in the slime.

Methods purporting to assess microbiological fouling by determining loss on ignition are meaningless. The enormous amount of dirt drawn into an open recirculating cooling system brings with it much organic material, including humus, humus precursors, wood fibers, and plant residues, all of which are mechanically trapped in and on various gelatinous deposits already described. Other limitations of the nonspecific loss on ignition have been discussed in another publication.[13]

Several years ago, in an attempt to resolve a debate on the nature of some deposits found on corrosion coupons from four different cooling systems, samples of each deposit were submitted to five different laboratories for independent analysis. Three of these laboratories were those of water service companies, the fourth was a research laboratory, and the fifth was an industrial laboratory. The microbiological analyses reported by each are shown in Table 8.1.

TABLE 8.1

Microbiological Analyses of Deposits

	Bacteria and Other Microorganisms			
Laboratory	System 1	System 2	System 3	System 4
A	None	1.7×10^6/g	10×10^6/g	12×10^6/g
B	10%	40%	10–20%	5–10%
C	Present	Present	Present	Present
D	None	None	None	None
E	26×10^6/g	20×10^6/g	76×10^6/g	14×10^6/g
Loss on Ignition, %	46	–	–	57

The debate continues.

When dry soil is blown into a cooling tower by the wind a fractionation of sorts takes place. The water-soluble components of the mixture that comprises humus go into solution and furnish nourishment for the resident microorganisms. The remainder of the humus mixture, along with the clay component, forms a colloidal solution, while the mineral sediment tends to accumulate on the bottom of the basin of the cooling tower. The amount of sediment laid down depends upon prevailing winds, the type of soil in the vicinity (although wind-borne dust can travel great distances), and the length of time between cleanings. As a specific example, however, consider a basin 66 ft. wide and 180 ft. long in a cooling tower serving a fluid catalytic cracking plant. At the end of two years of operation the plant was shut down for maintenance and the basin was found to contain a fairly uniform layer of mineral sediment approximately 6 in. deep. On the face of it this does not appear to be very much, but the volume of this sediment was $66 \times 180 \times 0.5$, or 5940 ft^3; its bulk density was 125 lb/ft^3. Thus, the basin contained 5940×125, or 742,500 lb of sediment accumulated during two years of operation, which means that, on the average, over 1000 lb of soil was taken in every day.

Generally speaking, the mineral sediment in a basin is relatively harmless, but it does harbor some unusual animal life rarely seen in other parts of a cooling system. Ciliates and rotifers are often present in large numbers. These are beneficial to some extent, as they are voracious predators of bacteria and other microorganisms. Members of both these classes attach themselves to particles or surfaces rather than swimming freely in the water. Rotifers do this by means of a forked foot supplied with glands that secrete an adhesive; sessile ciliates are usually secured by a stalk. Under the microscope the chewing action of the muscular pharynx of a rotifer can be seen clearly, as the organisms feeds on unicellular algae, bacteria, and organic debris. The basin sediment is a haven for anaerobic bacteria, which occasionally create a nuisance by releasing hydrogen sulfide, or by attacking the concrete basin. All of these organisms are more susceptible than bacteria to microbicides, but they are inaccessible within the sediment, where they thrive undisturbed.

Four methods for confirming the presence or absence of significant concentrations of microorganisms in fouling deposits are outlined below.

2.1 Direct Microscopic Examination

Transfer a loopful of fouling material to a clean glass slide, add a drop of distilled water, and cover with a glass cover slip. Using a 10 X eyepiece and a 10 X objective, focus on the slide and examine a number of fields noting the presence of oil droplets, wood fibers, filamentous algae, or any living organisms such as amoeba, paramecium, rotifers, and so forth. Also, bear in mind that the majority of microorganisms are symmetrical, whereas particles of dirt, clay, and other inorganics are, in general, irregularly shaped.

2.2 Examination After Staining

Transfer a loopful of fouling material to a clean glass slide, then proceed as directed in the procedure on page 180. If oil is present drench the slide in chloroform for two minutes after Step 3 on page 181, then continue with Step 4 and the rest of the procedure. Place the slide on the microscope stage and examine it using a 10 X ocular and a 40 X objective lens. As mentioned on page 180, 1 mg of bacteria spread over a slide yields about 80,000 bacteria per field.

2.3 Plate Count

Transfer 1 g of fouling material to a dilution tube prepared as in the procedure in Chapter IX, page 192. Mix thoroughly and prepare a series of dilutions as described on page 194, incubate for 48 h, count the number of colonies on the plate, and calculate the number of bacteria per gram of the original sample.

Suppose, for example, that the plate count of a particular fouling deposit is found to be 7.6×10^7 bacteria per gram. As 1.0×10^9 bacteria weigh 1 mg,[(4)] 7.6×10^7 bacteria weigh $76 \times 10^6/10^9$, or 0.076 mg. Therefore the concentration of living bacteria in the fouling deposit is 76 ppm, obviously an insignificant amount. It is sometimes claimed that although bacteria are few, the bulk of the material is still the product of their metabolism—in effect, waste products. The implication here would be that the bacteria in our example had produced metabolically over 13,000 times their own weight, an eminent impossibility.

Pursuing this example a little further, suppose that 1 mg of the material had been taken for staining in Section 2.2, above. Then 76,000 bacteria would have been distributed over the working area of the slide ($25 \times 50 = 1250$ mm^2), the number of bacteria per square millimeter would have been about 61, and the number per field using a 40 X objective would have been about 6 bacteria. Again, a negligible number compared to 80,000, the number that would be visible if the material were composed entirely of bacteria.

2.4 Determination of Total Protein

In Chapter X, Section 3, a procedure is given for determining protein. Bacteria contain 12–15 percent total protein, as measured by this procedure. Thus, if a particular fouling deposit were found to contain, say, 0.04 percent protein, this would indicate a bacterial content of approximately 0.3 percent. Protein concentration can also be related in a general way to viable plate counts.

The foregoing methods of examination provide positive identification of microbiological fouling.

References

(1) Chamot, E. M. and Mason, C. W. *Handbook of chemical microscopy.* vol. 1, 3rd ed. (1958), vol. 2 2nd ed. (1940) New York: Wiley.

(2) Baker, F. J. and Breach, M. R. 1967. *Handbook of bacteriological technique.* 2nd ed., London: Butterworths.

(3) American Public Health Assoc. *Standard methods for the examination of water and wastewater.* 14th ed., 1974, New York.

(4) Hughes, D. E. 1968. *Microbiology.* London: Institute of Petroleum.

(5) McKinney, R. E. 1962. *Microbiology for sanitary engineers.* New York: McGraw-Hill.

(6) Harrison, M. E. and Heukelekian, H. 1958. Slime infestation–literature review. I. *Sphaerotilis. Sewage Ind. Wastes* 30:1278.

(7) Orloff, H. S. and Raymore, H. B. 1962. *A book about soils.* New York: M. Barrows and Company.

(8) Oglesby, R. T., Christman, R. F., and Driver, C. H. 1967. The biotransformation of lignin to humus–facts and postulates. In *Advances in applied microbiology,* vol. 9, ed. W. W. Umbreit, New York and London: Academic Press.

(9) Alexander, M. 1965. Biodegradation: problems of molecular recalcitrance and microbial fallibility. In *Advances in applied microbiology,* vol. 7, ed. W. W. Umbreit, New York and London: Academic Press.

(10) U.S. Dept. of Agriculture Handbook No. 18. 1951. *Soil survey manual.* Washington, D.C.

(11) Van Olphen, H. 1963. *An introduction to clay colloid chemistry.* New York: Interscience.

(12) Weiser, H. B. 1949. *A textbook of colloid chemistry,* 2nd ed., New York: Wiley.

(13) McCoy, J. W. 1969. *Chemical analysis of industrial water.* New York: Chemical Publishing Company.

IX.

Microbiological Analysis

In this chapter four procedures for measuring the degree of microbiological contamination of cooling water are given, and a screening method for evaluating the relative toxicities of microbicides. A qualitative test for anaerobic sulfate-reducing bacteria is included in Section 1. In Section 2 a routine method is prescribed for estimating the total number of living bacteria in cooling water, by which means it is possible to evaluate semiquantitatively the effects of microbicides and to avoid wasting these expensive chemicals. The small initial expenditure for the equipment needed in this procedure is quickly recovered by the savings in costs of microbicides. Two special procedures are included in Sections 3 and 4 of this chapter for measuring the concentrations of autotrophic bacteria and fungi that are sometimes useful in the operation of cooling systems.

1. ANAEROBIC BACTERIA–QUALITATIVE TEST

Sulfate-reducing anaerobes, in particular *Desulfovibrio desulfuricans* and *Clostridia nigrificans*, often proliferate under slime or other accumulations attached to surfaces in dirty cooling systems as well as in basin sediments. As these organisms are suppressed by chromate[(1)] and by dissolved oxygen, they are not likely to be found in large numbers in the recirculating water in a chromate-treated cooling system. Griffin[(2)] reports, however, that about two years after switching from chromate to a nonchromate treatment with phosphonate and polymer, heavy concentrations of sulfate-reducing bacteria developed that caused severe corrosion in steel piping and in condenser heads and shells. For this reason it is advisable to check occasionally for anaerobes after adopting a nonchromate water treatment, or whenever a film of slime forms on the walls of the basin of a cooling tower.

Desulfovibrio and *Clostridia* metabolize sulfate, sulfite, and thiosulfate to hydrogen sulfide; the mechanism of their corrosive action on iron has been discussed on pp. 22-24 and in another publication.[(3)] The qualitative test in this section is based on the reaction of hydrogen sulfide with ferrous ion in the culture medium, sulfate API broth, which produces intensely black particles of ferrous sulfide. The complete formula of the medium is given on page 115.

A sample of cooling water is collected at the wall of the tower basin, and one milliliter of it is injected into a prepared vial of sulfate API broth. The vial is then placed in an incubator at 37 C and inspected daily for ten days. If black particles form in the broth within this time the presence of anaerobic sulfate-reducing bacteria is confirmed.

PREPARATION

Equipment:

1. Incubator, Thermolyne Constant Temperature (37 C). Sybron Corporation, P.O. Box 629, Rochester, New York 14602.
2. Glass Syringes, 2-½ ml, sterile, disposable. Becton, Dickinson and Company, Rutherford, New York.
3. Sulfate API Broth, prepared in 9-ml vials. Difco No. 0500-86, Difco Laboratories, Detroit, Michigan 48232.

PROCEDURE

1. Rub away any slime from a spot on the wall of the cooling tower basin, then fill a 2-½-ml sterile syringe with water at the wall.
2. Remove the aluminum seal from a vial of sulfate API broth. Force water out of the syringe so that the needle is full of water, then push the needle through the rubber septum in the neck of the vial and carefully introduce 1.0 ml of water.
3. Withdraw the needle, break it off, and discard the syringe.
4. Place the vial in an incubator at 37 C and examine the broth daily for the presence of black particles.
5. If black particles form within ten days sulfate-reducing bacteria are present. Discard the vial after ten days of incubation.

2. *VIABLE PLATE COUNT*

To measure the number of bacteria in cooling water, a 0.1-ml portion of water, or a dilution thereof, is spread uniformly over the surface of a prepared agar plate, which is then incubated at 30-37 C for 48 h. The number of colonies of bacteria that develop on the plate are counted, multiplied by the dilution factor, and reported as organisms per milliliter in the original sample. This method has been described in considerable detail in Chapter VI, so there is no need to repeat that discussion here, but a few words about the interpretation of results may be helpful.

Counts obtained by the following procedure are by no means an absolute measure of the total number of bacteria in a sample of water. The assumption is implicit that each colony that grows on an agar plate is made up of the progeny of a single bacterium, whereas, in reality, one colony is frequently produced by a collection of bacteria clinging to one another. Clumping is induced by cationic detergents and by other surface-active agents. Then too, after cell division a number of bacteria may remain attached to one another in chains, or trichomes, especially those that divide transversely, such as *Flavobacterium.*

A further complication in the interpretation of plate counts arises from the tendency of bacteria to adhere to charged surfaces. The following structural formulas of an amino acid represent a protein (and by extension, a bacterium) in three electrical configurations.

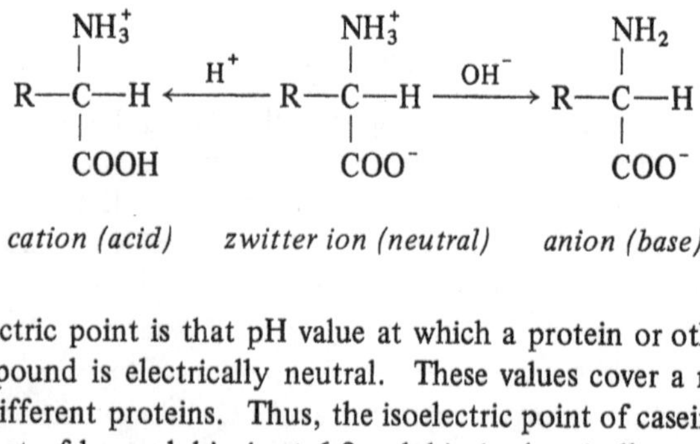

cation (acid) *zwitter ion (neutral)* *anion (base)*

The isoelectric point is that pH value at which a protein or other amphoteric compound is electrically neutral. These values cover a rather wide range in different proteins. Thus, the isoelectric point of casein is at a pH of 4.7; that of hemoglobin is at 6.8; globin is electrically neutral at 8.1. Because of this phenomenon the electrostatic adsorption of bacteria on clay, wood, steel, and inorganic compounds is strongly influenced by the pH of cooling water as well as by the concentration of salts dissolved in it. The adsorption of *Pseudomonas*, *Sarcina*, and *Bacillus* is strongest at pH 3-6. Higher pH values and increasing concentration of salts promote desorption of bacteria back into the cooling water.

Colonies may be transparent, translucent, or opaque; most bacteria in cooling water form opaque colonies, usually with a dull appearance. When examined under magnification they may be smooth, granular, or filamentous. Those with wrinkled edges are considered to be one of the two spore-formers *Clostridia* or *Bacillus.* Very small, almost pinpoint, colonies are seen for several days after treating a cooling system with chlorine dioxide as recommended in Chapter VI, indicating an inability of the organisms to grow. Colonies of capsule-forming bacteria on agar plates exhibit a characteristic moist, glistening surface described as mucoid. Some of these organisms, however, form capsules only under special conditions, notably in the presence of sugars. As the medium used for plate counts contains glucose there is certainly no assurance that bacteria that form mucoid colonies on agar plates are also able to produce slime in cooling water. A few soil bacteria, perhaps 15 percent of them, can generate gums, but again, they require sugar to do this.

Counting is easier if a Quebec colony counter[4,5] is used, although this device is not essential. Similarly, an autoclave is more convenient for sterilizing than a pressure cooker, but the latter will do very well if only a few plates are set each week. If freshly prepared agar plates are used confluent growths, or "spreaders," are frequently obtained caused by an excess of water on the surface of the agar. Plates should be inverted and stored in a refrigerator for two or three days before using. A similar growth may occur if the agar plate is inverted and placed in the incubator immediately after spreading. The plates should be allowed to stand for an hour before inverting and incubating.

PREPARATION

Equipment

1. Colony counter, Quebec, (Optional). American Optical Corporation, Buffalo, New York 14215.
2. Culture tubes, Pyrex, 16 × 150 mm, screw cap. Corning No. 9826, Corning Glass Works, Corning, New York 14830.
3. Petrie Dishes, plastic, sterile, 100 × 15 mm. Lab-Tek Products, Division of Miles Laboratories, Westmont, Illinois 60559.
4. Plate count agar: (a) prepared tubes, Difco No. 0479-37, or (b) dehydrated, Difco No. 0479-01. Difco Laboratories, Detroit, Michigan 48232.
5. Serological Pipettes, 1 in 1/10-ml, with plug. Falcon Plastics, 1950 Williams Drive, Oxnard, California 93030.
6. Spreaders–sterile glass rods shaped like hockey sticks.
7. Autoclave, Castle Model 999-C (Optional). Sybron Corporation, P.O. Box 629, Rochester, New York 14602.

8. Incubator, as in Section 1.
9. Veri-Whirl Mixer. Catalog No. 58810-006, Van Waters & Rogers Scientific, P.O. Box 3200, Rincon Annex, San Francisco 94119.

Pouring Agar Plates

I. *Using prepared agar:*

1. Place the required number of prepared agar tubes in a beaker of hot water with the water level just above the surface of the agar. Loosen the caps slightly and boil for 10 minutes on a hot plate.
2. Remove the beaker from the heat and let it cool to about 48 C (just warm to the touch). Tighten the caps and dry the outside of the tubes.
3. Raise the cover of a sterile Petrie dish just high enough at one edge to permit pouring the entire contents of a tube into the dish. Cover, and swirl gently to distribute the agar evenly.
4. Allow the agar to congeal, then store the plates inverted (lid down) in a refrigerator for 2-3 days before using.

II. *Using Dehydrated agar:*

1. Weigh 8.2 g of dehydrated agar and transfer it to a 500-ml Erlenmeyer flask, then add 350 ml of distilled water.
2. Place the flask in a 3-liter beaker containing about 1000 ml of water.
3. Transfer the beaker to a hot plate and boil until the agar is in solution.
4. Remove from the hot plate, cover the mouth of the flask with aluminum foil, and sterilize at 15-18 psi for 20 min in a pressure cooker, or autoclave.
5. Allow the agar to cool to about 48 C, then pour the plates directly from the flask. Do not attempt to measure the volume of agar for each plate, but estimate 9-10 ml. This amount covers about three-fourths of the bottom of a Petrie dish when poured, and after swirling, covers the bottom of the dish to a depth of 2-3 mm.
6. Allow the agar to congeal, then store the plates inverted in a refrigerator for 2-3 days before using.

Preparation of Dilution Tubes

1. Dissolve 9 g of NaCl in 1000 ml of distilled water.
2. Using a 10-ml Mohr pipette transfer 9.25 ml of the saline solu-

tion to each of a series of culture tubes and cap loosely.

3. Sterilize the tubes at 15–18 psi for 20 min. Approximately 0.25 ml of water is lost from each tube.
4. Cool, and tighten caps. Dilution tubes must be at room temperature when used.
5. When preparing used dilution tubes for reuse, wash them thoroughly in hot water, using a brush and detergent, then rinse them three or four times with distilled water.
6. Glass spreaders should be washed carefully, wrapped in aluminum foil, and sterilized along with the agar and dilution tubes.

Sterilizing with a Pressure Cooker

1. Add distilled water until its level is just below the rack in the bottom of the pressure cooker.
2. Place the items to be sterilized in the cooker, tighten the cover clamps, open the pressure release, and place the cooker on a hot plate.
3. Allow steam to escape through the pressure release valve for about 2 min, close the valve, and allow the pressure to increase to 15–18 psi (250–260 F). Maintain the pressure in this range for 20 min.
4. Remove the pressure cooker from the hot plate and allow it to cool until the pressure gauge reads zero. Do not open the pressure release valve until the pressure is zero or liquids in the cooker will boil over.
5. Open the pressure release valve and remove the contents of the cooker.
6. When using an autoclave follow the manufacturer's instructions.

PROCEDURE

Drawing Samples

1. Draw samples of cooling water for viable plate count in sterile, dry culture tubes from the basin or from the return line to the cooling tower.
2. Rinse the tube once or twice with sample, then fill it completely, and tighten the cap.
3. Return the sample to the laboratory, make dilutions, and spread plates within 2 h.

Making Dilutions

Note. Sterile, disposable, 1-ml serological pipettes are used. Packages of 20 should be opened by making a small slit in the end of the package opposite the tips. If the proper sequence is followed only one pipette is required regardless of the number of dilutions necessary for each sample. For routine control only dilutions of $1{:}10^4$ are necessary, as there is no concern with counts less than 100,000/ml.

1. Place three dilution tubes in a test tube rack and loosen the caps so they can be removed easily with the fingers of one hand.
2. Mix the sample thoroughly using a "Veri-Whirl" mixer, if available, to break up clumps of bacteria.
3. Transfer 1.0 ml of the mixed sample to the first dilution tube. Rinse the pipette several times by drawing the saline solution up and letting it flow down the inside of the pipette. Replace the cap and agitate the solution on the mixer.
4. Rinse the pipette two or three times with the first dilution, then transfer 1.0 ml of it to a second dilution tube. Rinse and mix as in Step 3.
5. Transfer 1.0 ml from the second dilution tube to a third dilution tube following the above steps, but leaving the pipette in the third tube.

Plating the Sample Dilution

1. Identify and date the covers of the plates to be spread. Make three plates from each dilution to be counted; for routine work only the 10^4 dilution (from the third tube) is needed.
2. Transfer to the center of each plate marked 10^4 a 0.1-ml aliquot of the solution from the third dilution tube.
3. If the 10^3 dilution is also to be counted rinse the pipette with the solution in the second dilution tube, then transfer 0.1 ml of this solution to each of the plates marked 10^3.
4. Spread the sample evenly over the surface of the agar with a sterile glass spreader. If more than one dilution is to be counted spread the plates of highest dilution first in order to use a single spreader.
5. Allow to stand 1–2 h after spreading, then invert the plates (cover down), and place them in the incubator for 48 h.
6. At the end of the incubation period remove the plates, count the number of colonies on each plate, average the results of each

dilution, and multiply the average by the dilution factor.

7. Report the number of bacteria per milliliter. Note that in the third dilution tube the sample has actually been diluted $1:10^3$, but only 0.1 ml of that dilution is spread on the plate, which introduces another factor of ten.

3. AUTOTROPHIC BACTERIA COUNT

Cooling water that is continuously contaminated with 20–50 ppm of ammonia more often than not will also contain 2–10 ppm of nitrite generated by the nitrifying bacteria *Nitrosomonas*, *Nitrosogloea*, and others in a process called nitrosification. Other genera, e.g., *Nitrobacter* and *Nitrocystis*, carry the oxidation on to nitrate, a reaction called nitrification. As these genera cannot use carbon compounds they do not grow on standard plate agar and so are not detected by the method in Section 2. The following procedure, although considerably more complicated than that in the previous section, does detect these bacteria. The inorganic medium takes sometime to prepare, and as the growth of autotrophes is poor on agar, silica gel must be used as the supporting medium for growth. Pramer(6) has studied the influence of various physical and chemical factors in the preparation of silica gel. One of the major difficulties in using this method is caused by the phenomenon of syneresis, which is the separation of water as the gel contracts during gelation. This water must be removed before spreading the plates, for only a small amount of moisture can cause confluent growth of nitrifying bacteria and thus prevent the development of discrete colonies, which even under the best of conditions are quite small.

Funk and Krulwich(7) have devised a method for preparing water-clear silica gels that are firm enough to be streaked with a glass spreader, as described in Section 2. The gel is prepared by aseptically combining equal volumes of the mineral medium and potassium silicate, followed by a small amount of phosphoric acid, which initiates gelation after about one minute. Plates must be poured quickly before gelation commences. The gels become firm enough to streak in about 15 min, but plates must be incubated in a moist atmosphere to prevent drying and cracking of the gels.

There is no reason to carry out the following procedure unless a cooling water consistently contains at least 100 ppm of ammonia.

PREPARATION

Equipment

1. As in Section 1 of this chapter.
2. Erlenmeyer flasks, 125 ml. Corning No. 5000, Corning Glass Works, Corning, New York 14830.

Reagents

1. Potassium silicate solution. Dissolve 20 g of silicic acid, H_2SiO_3, in 200 ml of distilled water containing 14 g of KOH, warming to hasten solution.
2. Phosphoric acid, 20 percent. Dilute 155 ml of 85 percent H_3PO_4 to 1000 ml, and mix thoroughly.
3. Potassium Dihydrogen Phosphate, 0.1 *M*. Dissolve 1.35 g of KH_2PO_4 in 100 ml of distilled water.
4. Autotrophic medium for nitrifying bacteria. Dissolve the following weights of salts in 90 ml of distilled water: $(NH_4)_2SO_4$, 0.7 g; $FeSO_4 \cdot 7H_2O$, 0.03 g; $MgSO_4 \cdot 7H_2O$, 0.15 g; NaCl, 0.3 g; $NaNO_2$, 0.5 g. Add to this solution 10 ml of 0.1 *M*-KH_2PO_4. Transfer the combined solutions to a 1000-ml volumetric flask, add 0.4 ml of trace elements solution, and dilute to 1000 ml. Add 2 g of $CaCO_3$, and mix thoroughly.
5. Trace elements solution. Prepare solutions containing 1000 ppm of each required element by dissolving the weight of salt listed in the second column of Table 9.1 in 100-ml portions of distilled water. Transfer the volumes of these solutions listed in the third column to a glass-stoppered mixing cylinder, dilute the mixture to 40 ml, and mix thoroughly. Add 0.4 ml of this solution to the mineral medium prepared above to give the approximate amounts of trace elements listed in the fourth column in Table 9.1.

Preparation of Silica Gel Plates

1. Add 20 ml of potassium silicate to a 125-ml Erlenmeyer flask and cover loosely with aluminum foil.
2. To another 125-ml flask add 20 ml of Autotrophic Medium, and to a third flask add 4 ml of 20 percent phosphoric acid, covering each with aluminum foil. Each set of the three solutions will ultimately make three plates. Prepare as many sets of three as required for immediate use; the poured plates do not keep well.

TABLE 9.1

Preparation of Trace Elements Solution

Element	*Compound*	g/100 ml	ml/40 ml	µg/1000 ml
Boron	H_3BO_3	0.572	2.1	21
Cobalt	$CoSO_4 \cdot 7H_2O$	0.477	1.4	14
Copper	$CuSO_4 \cdot 5H_2O$	0.391	1.7	17
Manganese	$MnSO_4 \cdot 2H_2O$	0.406	2.2	22
Zinc	$ZnSO_4 \cdot 7H_2O$	0.440	1.6	16

3. Transfer the Erlenmeyer flasks to an autoclave or pressure cooker and sterilize by heating for 15 min at 121 C.
4. Allow the solutions to cool, then pour the potassium silicate aseptically into the flask containing the Autotrophic Medium, and mix thoroughly by swirling.
5. Have three sterile Petrie dishes ready, then pour the mixture obtained in Step 4 into the flask containing 4 ml of 20 percent H_3PO_4. Swirl to mix and quickly pour the solution into the three plates, swirling each gently to distribute the gel evenly.
6. Gelation is complete in about 15 min after which the water of syneresis should be aseptically decanted.
7. Prepare dilutions as directed in the procedure for Viable Plate Count, page 193, and carefully spread 0.1 ml of an appropriate dilution over the surface of each plate. Allow the plates to stand for 30 min, then invert them.
8. Transfer the plates to an incubator at 25 C, stacking them inverted, and covering each stack with an inverted 1000-ml beaker to maintain a moist atmosphere.
9. Incubate for 72 h, then count the colonies using 5 X magnification, and report the number of autotrophic bacteria per milliliter as in the procedure for Viable Plate Count.

4. *FUNGUS COUNT*

The general term fungus encompasses molds (*Mucorales*), yeasts (*Saccharomyces*), rusts (*Uredinales*), mildews (*Erysiphaceae*), smuts (*Ustilagi-*

nales), and mushrooms (*Basidiomycetes*). Fungi that cause concern in cooling systems are those that instigate rot in wooden cooling towers, in particular *Basidiomycetes*, *Ascomycetes*, and *Fungi Imperfecti*. Others that cause plugging, e.g., some species of *Aspergillus*, are less frequently seen, but can be grown in colonies on maltose agar. Maltose, a disaccharide composed of two D-glucose molecules, is especially appropriate for the cultivation of fungi on account of its ease of fermentation and its acidic reaction; the latter encourages the growth of fungi while inhibiting that of bacteria. Deyhdrated Sabouraud maltose agar contains the following ingredients:

Neopeptone	10 g
Maltose	40 g
Agar	15 g

When hydrated the pH of the medium is 5.6. Neopeptone is a proprietary enzymatic digest that is an unusually good source of nitrogen for the development of fungi. Growth is rapid on Sabouraud maltose agar and uniform colonies of fungi are obtained. Although the growth of bacteria is inhibited at the low pH, some colonies of *Pseudomonas aeruginosa* may develop with their blue antibiotic pyocyanin pigmentation enhanced by this medium.

As the ambient air is often laden with spores of molds, yeasts, and other fungi, luxuriant and colorful growths are sometimes obtained when maltose plates are prepared from cooling water. The great majority of these are of no significance. The principal value of the fungus count is to cultivate molds causing some particular difficulty (perhaps plugging) in a cooling system so that the nuisance can be identified and eradicated. Thus, if profuse growth is obtained on the Sabouraud medium a biocide should be applied that is especially virulent to fungi–2,2-dibromo-3-nitrilopropionamide rather than methylene bisthiocyanate, for example.

PREPARATION

Equipment

As in Section 2 of this chapter.

Preparation of Agar Plates

1. Weigh 26 g of dehydrated Sabouraud Maltose Agar, Difco No. 0110-01, and transfer it to a 500-ml Erlenmeyer flask, then add

400 ml of cold distilled water.

2. Place the flask in a 3-liter beaker containing about 1000 ml of water.
3. Transfer the beaker to a hot plate and boil until the agar is in solution. Do not boil any longer than necessary as prolonged heating causes the gel to be too soft after pouring and gelling.
4. Remove from the hot plate, cover the mouth of the flask with aluminum foil, and sterilize at 15 psi for 15 min in an autoclave or pressure cooker.
5. Allow the agar to cool to about 48 C, then pour the plates directly from the flask.
6. Allow the agar to congeal, then store the plates inverted in a refrigerator for 24 h before using.

PROCEDURE

Drawing Samples

1. Draw samples of cooling water for fungus count from the return line to the cooling tower in dry, sterile culture tubes. As the purpose of the count is to determine whether the normal treatment is controlling fungal growth, samples should be drawn 24 h after applying a regular biocide treatment.
2. Rinse the tube once or twice with sample, then fill it completely, and tighten the cap.
3. Return the sample to the laboratory, make dilutions, and spread plates within 2 h.

Plating the Sample

1. Identify and date the covers of the plates to be spread. A single plate of the undiluted sample is usually sufficient.
2. Transfer a 0.1-ml portion of cooling water, or if necessary, a dilution thereof, to each plate, then spread the sample evenly over the surface of the agar with a sterile glass spreader.
3. Allow to stand 1–2 h after spreading, then invert the plates (cover down), and place them in an incubator at 37 C for 48 h.
4. At the end of the incubation period remove the plates and inspect them for growths of mold. If desired, count the colonies, but normally this is not necessary. The usual concentration of fungi in cooling water is 10–100 organisms per milliliter.

5. *EVALUATION OF TOXICANTS*

Toxicant evaluations are made by manufacturers to compare the effectiveness of their own and their competitors' microbicides. One of the ways to do this is to expose measured portions of a sample of cooling water containing 500,000 to a few million bacteria per milliliter, to given concentrations of biocides for a certain number of hours. The number of viable bacteria in the various portions of water is reduced to an extent dependent on the concentration and toxicity of the biocide. At the end of the selected period of exposure viable plate counts are made and the population density relative to the undiluted sample is calculated, usually as "percent kill," although the "percent surviving" is the significant number. Thus, if the original water contained 1.5×10^6 bacteria per milliliter, and after treating with a biocide, 99 percent had been killed, there would still be 15,000 survivors per milliliter that would continue to multiply.

The same procedure is used for formulations intended as algicides, for in most if not all instances, biocides effective against bacteria also kill algae. Furthermore, this test is much less time-consuming than those using free-flowing algae. It is also pertinent to note that bacteria exist wherever algae are found, although the reverse does not hold true, especially in rapidly flowing water.

There is seldom any occasion for a customer to carry out a toxicant evaluation, for with the exceptions discussed on pages 135–142, it is a laboratory procedure yielding results that have limited value for predicting those that are obtained in an operating cooling system. Vendors use these, however, to promote sales, so it is of interest to outline the procedure, which is as follows:

1. Transfer 100-ml portions of a sample of cooling water containing 5.0×10^5–3.0×10^6 bacteria per milliliter to a series of 250-ml Erlenmeyer flasks.
2. Prepare stock solutions containing 1 g/100 ml (10,000 ppm) of the biocides to be evaluated, in either water or methanol, as appropriate.
3. Add biocide to the water in the Erlenmeyer flasks to give the desired concentration, e.g., for 30 ppm add 0.30 ml. Control flasks should receive an equal amount of methanol if alcohol is used as a solvent in the biocide formulation.
4. Allow the solutions to stand with occasional agitation for a

selected period of time, usually 3, 6, and 24 h, then remove a sample with a sterile pipette, and determine the plate count as described in Section 2.

5. Calcualte percent kill (if you are a seller), or percent survivors (if you are a buyer), based on the average plate count of the control samples.

Toxicant evaluations of chlorine, bis(tri-n-butyltin)oxide, dodecylguanidine hydrochloride, and acrolein are listed in Table 9.2 in the form most often used by vendors of biocides. These results were obtained using a sample of chromate-treated cooling water containing 1.1×10^6 bacteria per milliliter:

Table 9.2 shows, among other things, that 10 ppm of 35 percent dodecylguanidine hydrochloride killed over 99 percent of the bacteria in 3- and 6-h exposures, but that in 24 h only 8 percent were killed. These figures mean that within 24 h the few survivors of the early hours had multiplied to 92 percent of their original concentration. It also suggests that if an actual cooling system were treated with only 10 ppm of dodecylguanidine hydrochloride, within a few days there would probably be more bacteria in the water than before the treatment. In effect, the biocide in deficient concentration stimulates growth, as discussed in Chapter VI.

According to these evaluations the order of toxicities and the effective concentrations of the four biocides are: chlorine, 0.5 ppm; bis(tributyltin) oxide, 25 ppm; dodecylguanidine hydrochloride, 30 ppm; acrolein, ineffective at the concentrations tested. Finally, it is emphasized again that the values in Table 9.2 record what happened in Erlenmeyer flasks, not necessarily the results that would have been obtained in a cooling system.

TABLE 9.2

Toxicant Evaluations (as Percent Kill)

Toxicant:	*Chlorine (100%)*			*Bis(tributyltin) Oxide (5 %)*			*Dodecylguanidine (35 %)*			*Acrolein (100%)*	
ppm	3 h	6 h	24 h	3 h	6 h	24 h	3 h	6 h	24 h	3 h	24 h
0.2	38	23	10	–	–	–	–	–	–	13	22
0.5	99+	99+	99+	–	–	–	–	–	–	44	53
2	99+	99+	99+	79	85	31	–	–	–	52	71
5	99+	99+	99+	99+	99+	65	43	33	7	58	85
10	99+	99+	99+	99+	99+	93	99+	99+	8	76	89
25	–	–	–	99+	99+	99+	99+	99+	92	–	–
30	–	–	–	99+	99+	99+	99+	99+	99+	–	–

References

(1) Drummond, J. P. M. and Postgate, J. R. 1949. A note on the enumeration of sulfate-reducing bacteria in polluted water and on their inhibition by chromate. *J. Appl. Bacteriol.* 18:307.

(2) Griffin, R. W. Nov. 11, 1974. Water management saves money in refinery cooling systems. *Oil and Gas J.* 72:198.

(3) McCoy, J. W. 1974. *The chemical treatment of cooling water.* New York: Chemical Publishing Co.

(4) Archambault, J., Curot, J., and McGrady, MacH. 1937. The need of uniformity of conditions for counting plates. *Am. J. Public Health* 27:809.

(5) Richards, O. W. and Heijn, P. C. Sept.-Oct., 1945. An improved Darkfield Quebec colony counter. *J. Milk Technol.* 8 (5), 253.

(6) Pramer, D. 1957. The influence of physical and chemical factors on the preparation of silica gel media. *Appl. Microbiol.* 5:392.

(7) Funk, H. B. and Krulwich, T. A. 1964. Preparation of clear silica gels that can be streaked. *J. Bacteriol.* 88:1200.

X.

Chemical Methods of Analysis

Several chemical methods are described in this chapter that are useful in the microbiological control of recirculating cooling systems. In Section 1 directions are given for determining several biocides often used in cooling water. These include chlorine, dodecylguanidine hydrochloride, quaternary ammonium salts, chlorophenols, and 2-mercaptobenzothiazole. Although a method has been published for determining methylene bisthiocyanate,[(1)] it is used at such low concentrations in cooling water (3-4 ppm), and disappears so rapidly that there is seldom any practical purpose in attempting to measure it. In Section 2 a method is provided for estimating mono- and diethanolamine, both of which cause serious microbiological fouling when they leak into cooling water. As ammonia interferes in the determination of the alkanolamines, a method is also provided for measuring it. As mentioned in Chapter VIII, there is a strong bias in the water-treating field toward characterizing every gelatinous fouling deposit as microbiological. For those who would rather know than speculate about this, a procedure is given in Section 3 for determining total protein in fouling material. Finally, in Section 4 a method is described for assaying concentrated formulations of cationic surfactants.

1. DETERMINATION OF SELECTED BIOCIDES

The procedures in this section are intended to be used for measuirng the concentration of some common biocides in treated cooling water. Normal concentrations of corrosion inhibitors and antiscalants such as chromate, zinc, phosphates, phophonates and polyacrylates do not interfere.

1.1 Chlorine

Chlorine is widely used as a microbicide, but it is incompatible with a number of water-treating chemicals including aminomethylenephosphonate, ferrocyanide, lignins, 2-mercaptobenzothiazole, and polyacrylamide. When chlorination is used the dosage should be such that the "total residual chlorine," as measured by reaction with ortho-tolidine reagent for 3 min, is not more than 2 ppm chlorine, while the "free available chlorine," as measured by reaction with the same reagent for 5 s is 0.5–0.8 ppm. If the water treatment is compatible with chlorination the free available chlorine recommended should be attained within 1-2 h, after which it should be maintained at 0.5–0.8 ppm for an additional hour.

Chlorine rapidly oxidizes ortho-tolidine (3,3′-dimethylbenzidine) to the corresponding holoquinone, which is intensely yellow:

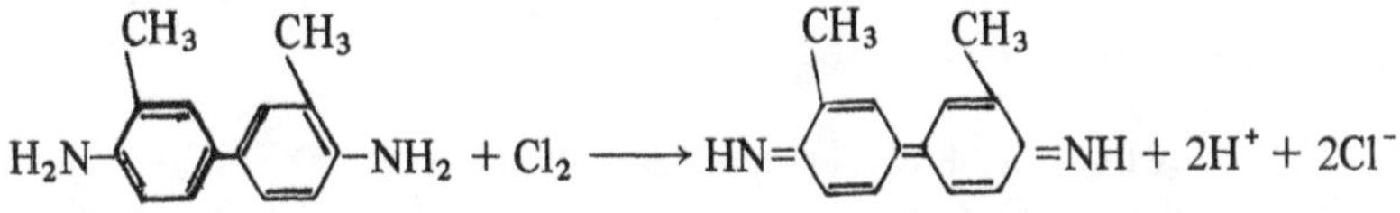

(10.1)

This provides a sensitive test for available chlorine, but the fugitive nature of this powerful oxidizing agent requires that testing for chlorine be done at the cooling tower. Several different test kits are on the market containing test tubes, permanent color standards, reagent solution, and a color comparator of some sort. Records should be kept of available and residual chlorine, times of sampling, feed rate of chlorinator, and duration of chlorination.

PREPARATION

Equipment

Chlorine block comparator. Range 0.2–2.0 ppm chlorine. Available from LaMotte Chemical Products Company, Chestertown, Maryland 21620.

Reagent

Ortho-tolidine reagent. Dissolve 1.35 g of o-tolidine dihydrochloride (Eastman No. 250) in 500 ml of water and 350 ml of HCl. Store in a dark bottle, protected from sunlight, and renew after 6 months.

PROCEDURE

Free Available Chlorine

1. Chill a sample of cooling water for 2–3 min in ice water.
2. Add the chilled sample to one of the test tubes provided in the kit, filling it to the mark.

3. Rapidly add 8 drops of o-tolidine reagent to the test tube, and mix.
4. Place the test tube in the comparator block, match the color of the solution with the nearest permanent color standard, and read within 5 s after adding the o-tolidine solution.

Total Residual Chlorine

Proceed as above, but do not chill the sample, and allow the color to develop for 3 min before matching with the nearest standard.

1.2 Cationic Surfactants

The following procedure is based on a method devised by Few and Ottewill[2] for determining small concentrations of cationic detergents. It is applicable to any of the salts of dodecylguanidine, various quaternary ammonium salts, and other cationic surfactants. A sample of cooling water containing a cationic biocide is treated with Orange II in the presence of a buffer so that yellow, chloroform-soluble complexes are formed between the dye and cationic biocides. The colored complex is extracted with chloroform and its intensity is measured in a spectrophotometer.

It was mentioned previously that the concentration of dodecylguanidine and quaternary ammonium salts decreases rapidly in cooling water because of the formation of electrostatic bonds between cationic compounds and the profusion of negative charges on wood, dirt, inorganic oxides, and microbiological debris in cooling systems. The quantitative effect of this is shown in Table 10.1. The values in the second column were those obtained at specified intervals after adding the equivalent of 32 ppm of a 35 percent solution of dodecylguanidine hydrochloride. The values in the third column are the concentrations that would have been present under normal depletion. The total volume of the cooling system, V, is 300,000 gal and the rate of blowdown is 3180 gal/h.

As a rule, it is more convenient to prepare the calibration curve in terms of total formulation rather than the active ingredient. The following procedure is written to apply to a 35 percent solution of dodecylguanidine hydrochloride, but it can easily be adapted to measure the usual 50 percent formulations of quaternary ammonium salts.

PREPARATION

Reagents

1. Standard solution of formulation containing dodecylguanidine

TABLE 10.1

Depletion Rate of Dodecylguanidine Hydrochloride

Minutes after addition of 32 ppm	*ppm Dodecylguanidine Hydrochloride*	
	Actually measured	Normal Depletion Rate
15	20	31.9
30	18	31.8
60	14	31.7
90	12	31.5
120	11	31.3
180	11	31.0
240	7	30.7
12 h	0.8	26.4
24 h	0.4	24.8

hydrochloride or other cationic surfactant. Accurately weigh 1 g of biocide, dissolve it in water, and dilute to 1000 ml in a volumetric flask (1 ml = 1 mg).

2. Buffer solution. Dissolve 20 g of NaCl and 2 g of KH_2PO_4 in 1000 ml of water and adjust the pH of the solution to 5.5-6.0 with dilute sodium hydroxide.
3. Orange II solution. Dissolve 0.1 g of p-(β-naphthol-azo)benzenesulfonic acid (Eastman No. 196) in 1000 ml of distilled water containing 5.8 g of NaCl. Extract this solution with successive 50-ml portions of chloroform until the extracts are colorless.

Standardization

1. Dilute the prepared standard solution ten-fold to produce a solution containing 0.10 mg/ml, and transfer 2.0, 4.0, 8.0, and 10.0 ml of the dilute standard to a series of 250-ml separatory funnels equipped with Teflon stopcocks, and each containing 100 ml of the buffer solution. Include a blank.
2. Let stand 5-10 min, then add 15 ml of Orange II reagent, mix, and again let stand for 5-10 min.
3. Using a pipette, transfer 50.0 ml of chloroform to each separatory funnel, and shake vigorously for 1 min.
4. Allow the layers to separate, then filter the chloroform layer

through a plug of glass wool in the stem of the funnel, collecting enough extract to fill a 1-cm cuvette.

5. Measure the transmittances of the extracts at 480 mμ, and plot the transmittances obtained against milligrams of formulation on semi-logarithmic graph paper. The blank is used to zero the instrument.

PROCEDURE

1. Transfer a suitable portion of cooling water (25–50 ml) to a 250-ml separatory funnel containing 100 ml of buffer solution, and let stand 5–10 min.
2. Add 15 ml of Orange II reagent and let stand an additional 5–10 min.
3. Transfer 50.0 ml of chloroform to the separatory funnel, and shake vigorously for 1 min.
4. Allow the layers to separate, and filter the $CHCl_3$ layer through a plug of glass wool in the stem of the funnel into a dry 1-cm cuvette.
5. Measure the transmittance at 480 mμ, and read the corresponding milligrams of formulation from the prepared calibration curve.

$$\text{ppm formulation} = \frac{\text{mg formulation} \times 10^6}{\text{ml sample} \times 10^3}$$

1.3 Pentachlorophenol

Solutions of sodium pentachlorophenate at concentrations of 2000–3000 ppm are recommended(3) for spraying the lumber in existing cooling towers as protection against fungus invasion of the wood. As this chemical is extremely persistent and recalcitrant to biological oxidation, it is important to limit the amount that enters a biological waste treatment facility. This can be done by spraying one cell at a time and keeping track of the amount of pentachlorophenol dissolved in the recirculating water. This is related to the size of the spray nozzle: a large nozzle used in a small cooling tower may produce concentrations of 40–60 ppm of sodium pentachlorophenate in the recirculating water, which is excessive; smaller nozzles reduce the contamination to 10–15 ppm. The following procedure using 4-aminoantipyrine is a modification of one proposed by Goto, et al.(4)

PREPARATION

Reagents

1. 4-Aminoantipyrine reagent. Dissolve 0.2 g of 4-aminoantipyrine (Eastman No. 6902) in 100 ml of distilled water.
2. Disodium phosphate solution. Dissolve 94 g of Na_2HPO_4 in 1000 ml of distilled water.
3. Ferricyanide solution. Dissolve 10 g of $K_3Fe(CN)_6$ in 100 ml of distilled water.
4. Standard Pentachlorophenate Solution. Dissolve 1.00 g of Dowicide G-ST, Dow Chemical Company, Midland, Michigan 48640, in 1000 ml of distilled water (1 ml = 1.00 mg).

Standardization

1. Dilute the prepared standard solution 100-fold to produce a solution containing 10 μg/ml.
2. Transfer 2.0, 6.0, 14.0, and 20.0 ml of the dilute standard solution to a series of 125-ml separatory funnels, and dilute each to 25 ml, including a blank.
3. Using a pipette, transfer 20.0 ml of reagent grade mixed xylenes to each funnel, add 2 ml of HCl to each, and shake for 5 min.
4. Allow the layers to separate, then with a pipette transfer 10 ml of the xylene layer in each funnel to a second 125-ml separatory funnel.
5. Add 10 ml of disodium phosphate solution and 5 ml of 4-aminoantipyrene reagent to each organic portion, shake 30 s, then let stand 3 min.
6. Add 1.0 ml of potassium ferricyanide solution, shake 30 s, and let stand 15 min.
7. Transfer a portion of the organic layer to a 1-cm cuvette and measure its transmittance at 574 mμ.
8. Plot the transmittances obtained against the corresponding micrograms of sodium pentachlorophenate on semi-logarithmic graph paper. Note that the volumes taken correspond to 10, 30, 70, and 100 μg in 10 ml of xylene.

PROCEDURE

1. Extract a suitable sample of cooling water acidified with 2 ml of

HCl with 20 ml of reagent grade mixed xylenes.

2. Proceed as in *Standardization* starting with Step 4 and continuing through Step 7.
3. From the prepared calibration curve read the number of micrograms corresponding to the transmittance obtained. This number multiplied by 2 is the number of micrograms in the original sample.

$$\text{Sodium Pentachlorophenate (ppm)} = \frac{\mu g \times 2}{\text{ml sample}}$$

1.4 2-Mercaptobenzothiazole

The sodium salt of 2-mercaptobenzothiazole is used in some formulations of biocides and also as a corrosion inhibitor in many nonchromate treatments. It is especially efficient for protecting copper alloys, as the sulfur atom forms a strong chemisorption bond with metallic copper. The material is also called 2-thiazole-thiol and 2-benzothiazolethiol.

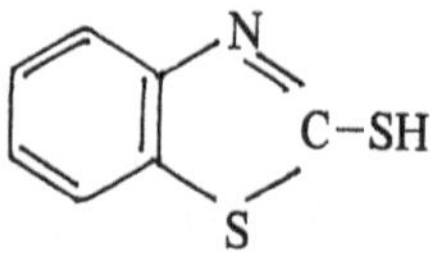

The compound is not compatible with chlorination, as it is oxidized by strong oxidizing agents (peroxide, hypochlorite) to the inactive (and insoluble) disulfide.

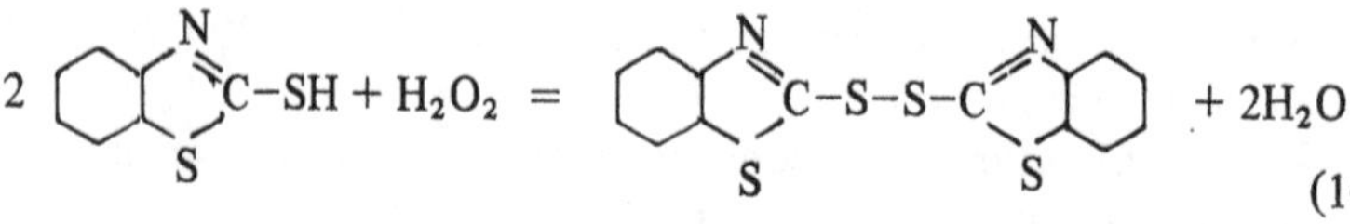

(10.2)

The concentration of sodium 2-mercaptobenzothiazole in cooling water is normally in the range of 1-2 ppm, and its concentration in water-treating formulations ranges from 6-10 percent. A common property shared by all aromatic compounds is the chemical's ability to absorb ultraviolet radiation. The maximum absorption of the sodium salt in aqueous solution is at 310 mμ with a secondary absorption at 230 mμ. If the free acid is dissolved in chloroform the maximum absorption shifts to 329 mμ. In the following procedure the sample of cooling water is acidified, and the free acid form is extracted with chloroform to increase the sensitivity of the determination. In general, it is convenient to prepare the calibration curve in terms of the whole formulation.

PREPARATION

Reagents

Standard solution of formulation containing sodium 2-mercaptobenzothiazole. Accurately weigh 1 g of the water-treating formulation, dissolve it in water, and dilute to 1000 ml in a volumetric flask (1 ml = 1 mg).

Standardization

1. Dilute the prepared standard solution ten-fold to produce a solution containing 0.10 mg/ml, and transfer 1.0, 2.0, 3.0, 4.0, and 5.0 ml to a series of 250-ml separatory funnels containing about 50 ml of water, and equipped with Teflon stopcocks.
2. Add 2 drops of 1:1 HCl to each funnel, and mix thoroughly.
3. With a pipette transfer 10.0 ml of $CHCl_3$ to each funnel, stopper, and shake the funnel for 3 min.
4. Allow the layers to separate, then filter the chloroform layer through a plug of glass wool in the stem of the funnel into a dry 1-cm cuvette.
5. Measure the transmittances at 329 mμ, and plot the transmittances obtained against milligrams of formulation on semi-logarithmic graph paper.

PROCEDURE

1. Transfer a 50-ml sample of cooling water, filtered if not clear, to a 250-ml separatory funnel equipped with a Teflon stopcock.
2. Add 2 drops of 1:1 HCl, and mix.
3. With a pipette transfer 10.0 ml of $CHCl_3$ to the separatory funnel, stopper, and shake for 3 min.
4. Allow the layers to separate and filter the $CHCl_3$ layer through a plug of glass wool in the stem of the funnel into a dry 1-cm cuvette.
5. Measure the transmittance at 329 mμ and read the corresponding milligrams of formulation from the prepared calibration curve. A blank should be run each day and deducted from the result obtained for the sample.

$$\text{ppm formulation} = \frac{\text{mg formulation} \times 10^6}{\text{ml sample} \times 10^3}$$

2. DETERMINATION OF ALKANOLAMINES AND AMMONIA

2.1 Alkanolamines

Monoethanolamine and diethanolamine leaking into cooling water cause so much trouble that it is desirable to have an early warning of their presence to avoid the growth of mold and bacteria in contaminated cooling water. This subject has been discussed at some length in Chapter VII, pp. 162 and pp. 173–175 so it will not be repeated here.

Ammonia in concentrations in excess of 25 ppm interferes in the following procedure; in this event it is necessary to determine total ammonia by the Kjeldahl method and subtract free ammonia as determined in Subsection 2.2, below. The method for alkanolamines is not precise; it is intended to show the presence of rather gross contamination of cooling water by mono- or diethanolamine. The procedure is written for monoethanolamine, but exactly the same directions are applicable for measuring diethanolamine. If chromate is used as a corrosion inhibitor approximately the same concentration should be used in the blank.

PREPARATION

Reagents

1. Standard monoethanolamine solution. Accurately weigh 0.50 g of monoethanolamine (Eastman No. 13074), dissolve it in water, and dilute to 1000 ml in a volumetric flask (1 ml = 0.50 mg).
2. Sodium hydroxide, ≈ 1 *N*. Dissolve 4 g of NaOH in 100 ml of distilled water.
3. Colorimetric reagent, 1, 2-Naphthoquinone-4-sulfonic acid, sodium salt. Baker No. R639, J. T. Baker Chemical Company, Phillipsburg, New Jersey 08865. Immediately before using, dissolve 0.3 g of the reagent salt in 10 ml of distilled water in a 30-ml beaker. Warm gently and stir continuously until dissolved. This reagent must be used within 30 min of preparation.

Standardization

1. To a series of 50-ml glass-stoppered mixing cylinders add 40 ml

of water and 1.0, 2.0, 3.0, and 4.0 ml of the standard solution of monoethanolamine corresponding to 0.5, 1.0, 1.5, and 2.0 mg of MEA. Include a blank.

2. Dilute each cylinder to 50 ml and mix thoroughly.
3. To each cylinder add 3 drops of 1 *N*-NaOH, and mix, then add 0.2 ml of colorimetric reagent, mix again, and let stand for 20 min.
4. Measure the transmittance of each solution at 480 mμ, and plot the transmittances obtained against milligrams of monoethanolamine on semi-logarithmic graph paper. If desired, a similar calibration can be prepared for diethanolamine.

PROCEDURE

1. Transfer an appopriate sample to a 50-ml glass-stoppered mixing cylinder, and dilute to 50 ml with water. (If the sample contains chromate, dilute with water containing approximately the same concentration of chromate.)
2. Add 3 drops of 1 *N*-NaOH, and mix, then add 0.2 ml of colorimetric reagent, mix again, and let stand for 20 min.
3. Measure the transmittance at 480 mμ and read the corresponding milligrams of alkanolamine from the prepared calibration curve.

$$\text{MEA (ppm)} = \frac{\text{mg} \times 10^6}{\text{sample (ml)} \times 10^3}$$

2.2 Ammonia

PREPARATION

Reagents

1. Nessler's reagent. Dissolve 34.9 g of KI in 100 ml of water, and gradually add 45.5 g of HgI_2, stirring until solution is complete. Dissolve 112 g of KOH in 200 ml of water, and cool to room temperature. Pour the two solutions together, and dilute to 1000 ml with water. Allow the solution to stand 2-3 days before using to settle the small amount of insoluble material that is present.
2. Standard Ammonia Solution. Dissolve 0.314 g of NH_4Cl in 1000 ml of water (1 ml = 0.1 mg).

Standardization

1. Transfer 1.0, 2.0, 3.0, 4.0, and 5.0 ml of the standard ammonia solution to a series of volumetric flasks, and include a blank.
2. Dilute the solution in each flask to about 75 ml, then add 5.0 ml of Nessler's reagent to each flask while swirling. Dilute to volume and mix well.
3. Let the solutions stand 30 min, then determine the transmittance of each solution at 420 mμ against the blank solution using 1-cm cuvettes.
4. Plot the milligrams of NH_3 against the corresponding transmittances on semi-logarithmic graph paper.

PROCEDURE

1. Transfer a measured sample of cooling water to a 100-ml volumetric flask.
2. Add 5.0 ml of Nessler's reagent to the sample solution, dilute to volume, and mix well.
3. Allow the solution to stand 30 min, then measure its transmittance at 420 mμ against a portion of cooling water as a blank, using 1-cm cuvettes.
4. From the prepared calibration curve read the corresponding milligrams of ammonia.

$$NH_3\ (\text{ppm}) = \frac{\text{mg } NH_3 \times 10^6}{\text{ml sample} \times 10^3}$$

3. DETERMINATION OF PROTEIN IN FOULING DEPOSITS

In the absence of a microscope or suitable magnifier it is possible to estimate the concentration of bacteria in a fouling deposit by determining total protein. Lowry, et al.(5) developed a colorimetric method for determining protein that was later adapted by Hill, et al.(6) for estimating the concentration of microorganisms in petroleum products, after they noted that spread plates, membrane filtration, and centrifugation methods failed to yield consistent maximum counts of organisms known to be present. They did find, however, a reasonable correlation between plate counts of *Escherichia coli* and the Lowry method. A calibration for the latter is prepared from standard solutions of bovine albumin, as described on page 216 under *Standardization.* Hill, et al.[6] reported the correlation in Table 10.2; the mg of bacterial/ml is obtained by dividing the plate count by 10^9, the approximate number of bacteria per milligram.

TABLE 10.2

Relationship Between Plate Counts and Total Protein

μg Bovine Albumin/ml	*SPC E. coli/ml*	*mg Bacteria/ml*
12	50×10^6	0.050
24	96×10^6	0.096
38	150×10^6	0.150
53	205×10^6	0.200
74	268×10^6	0.268
91	336×10^6	0.336

Protein constitutes 12–15 percent of the fresh weight of many soil bacteria and algae, or about 50 percent of their dry weight. Thus, by measuring total protein in a fouling deposit it is possible to make a rough estimate of the total weight of bacteria in the deposit, as explained on page 186. Folin and Ciocalteu[7] originally used the Folin phenol reagent to measure tyrosine and tryptophan in proteins; Lowry, et al.[5] later modified the method to improve its sensitivity and reproducibility. The formation of color results from a combination of two reactions: (a) the reaction of cupric ion with protein in alkaline solution; (b) the reduction of phosphomolybdic-phosphotungstic acid reagent by the copper-treated protein. Copper increases the intensity of the final color by as much as fifteen-fold.

Biuret and its derivatives,[8] as well as polypeptides and protein, form pink coordination complexes with cupric ion in alkaline solution (reaction (a), above). Also, the copper solution dissolves many proteins readily. The intense yellow color of Folin-Ciocalteu reagent disappears shortly after the solution is added, presumably because of the dissociation of phosphomolybdate. This reagent must be added rapidly with continuous mixing to develop the maximum final blue color. The bovine albumin standard used to prepare the calibration curve must be freshly prepared immediately before use, as albumin tends to undergo surface denaturation. These solutions are also deteriorated rapidly by the action of bacteria.

Before proceeding with the rather complicated quantitative chemical method it is advisable to make a qualitative check for the presence of protein and nucleic acids. To do this, transfer a portion of the suspect fouling material to a dilution tube, add water, then transfer the tube to a sonic oscillator. This will disrupt any microbiological cells present, releasing protein and nucleic acids into the solution. Centrifuge briefly, if necessary, then transfer a portion of the clear solution to a spectrophotometer cuvette, and check for characteristic absorbance at 260 and 280 mμ. If there is a positive response, a quantitative determination of protein can be made, if desired. If there is no response, a negligible amount of microbiological debris is present in the fouling material.

PREPARATION

Reagents

Solution No. 1. Dissolve 4 g of Na_2CO_3 in 50 ml of water, add 20 ml of 1 *N*-NaOH, and dilute to 200 ml with water.

Solution No. 2. Dissolve 1 g of $CuSO_4 \cdot 5H_2O$ in 50 ml of water, add 50 ml of a 4 percent solution of sodium tartrate, and dilute to 200 ml with water. Prepare this reagent just before it is to be used.

Solution No. 3. Add 2.0 ml of Solution No. 2 to 100 ml of Solution No. 1, and mix thoroughly.

Solution No. 4. Folin-Ciocalteu Phenol Reagent, approximately 2 *N*. Catalogue No. 2690, Hartman-Leddon Co., 60th and Woodland Avenue, Philadelphia, Pennsylvania 19143. Dilute the purchased reagent with an equal volume of water just before using.

Standard Protein Solution. Dissolve 0.100 g of crystallized bovine plasma albumin (Stock No. 2266-01, Reheis Chemical Co., Division of Armour Pharmaceutical Co., Chicago, Illinois 60690) in 25-30 ml of water in a small beaker. Transfer the solution to a 1000-ml volumetric flask, dilute to volume, and mix thoroughly. This solution foams considerably, but the foam collapses on standing (1 ml = 0.10 mg).

Standardization

1. To a series of 50-ml volumetric flasks add 5.0, 10.0, 15.0, 20.0, and 25.0 ml of the standard protein solution, corresponding to 0.50, 1.0, 1.5, 2.0, and 2.5 mg of protein, respectively.
2. Adjust the total volume in each flask to 25 ml, and include a blank.
3. To each flask add, while swirling, 15 ml of Solution No. 3, mix

thoroughly, and allow to stand exactly 10 min.

4. While swirling add, all at once, 3 ml of Solution No. 4, dilute to volume, mix, and allow to stand exactly 30 min.
5. Using 1-cm cuvettes, measure the transmittances of the prepared solutions at 720 mμ, and plot the results against the corresponding milligrams of protein on semi-logarithmic graph paper.

PROCEDURE

1. Transfer 1.0 g of fouling material to a dilution tube containing saline solution prepared as described on page 193, and mix thoroughly using a "Veri-Whirl" mixer or sonic oscillator, if available.
2. Allow insoluble material to settle, then transfer a suitable aliquot, usually 1.0 ml, of the supernatant liquid to a 50-ml volumetric flask.
3. Adjust the volume to 25 ml with water, and continue as directed under *Standardization*, starting with Step 3.

$$\text{Percent Protein} = \frac{\text{Protein (mg)} \times 100}{\text{sample (g)} \times (\text{aliquot}/10) \times 1000}$$

4. *ASSAY OF CATIONIC SURFACTANT FORMULATIONS*

Most methods for assaying formulations of cationic surfactants are based on the formation of insoluble salts with various anions. Hager, et al.[9] have devised a procedure based on the insolubility of the triiodides of quaternary ammonium compounds, and Flotow[10] has given several methods, including precipitation as dichromate, which is prescribed here.

The molecular weight of the cationic salt to be determined must be known. Quaternary ammonium salts prepared from dodecylbenzene derived from petroleum[11] have molecular weights a little higher than calculated because of the presence of higher homologs. For instance, the calculated molecular weight of N-dodecylbenzyl-N,N,N-trimethylammonium chloride is 354, while the empirical weight is about 358; the calculated molecular weight of N-dodecylbenzyl-N,N,N-triethylammonium chloride is 396, and the empirical weight is about 400. These differences, however, are insignificant in the analysis of commercial products sold as concentrated aqueous solutions. The triethyl compound is taken as an example in the following discussion of the procedure.

A measured amount of standard potassium dichromate is added to a weighed sample of the quaternary formulation, and the insoluble quaternary dichromate precipitates.

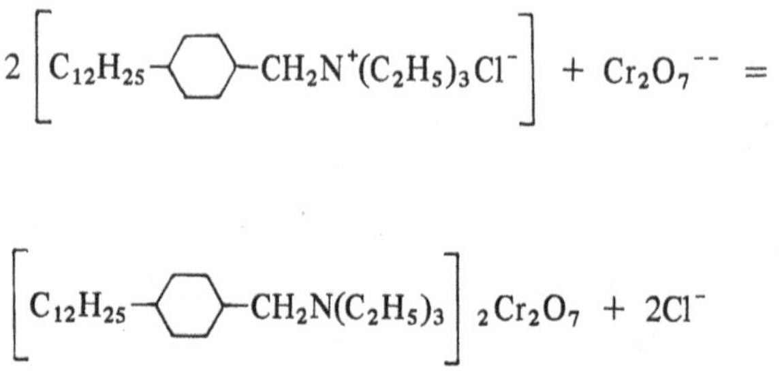

(10.3)

The precipitate is filtered, and the excess of dichromate is determined by adding iodide and titrating the resulting iodine with standard thiosulfate. The derivation of this rather complicated calculation is as follows:

A measured portion of potassium dichromate solution is titrated with standard thiosulfate to give the "blank." The same volume of dichromate solution is added to the weighed sample, part of it is precipitated, and the excess is titrated to give the "titration." The number of equivalents of dichromate precipitated is given by the expression

$$\frac{(\text{Blk} - \text{titr})(\text{ml}) \times N\ S_2O_3^{--}(\text{eq}/l)}{1000\ (\text{ml}/l)}$$

This expression multiplied by the equivalent weight of dichromate gives the number of grams of dichromate precipitated. The figure obtained is then divided by the grams of dichromate precipitated by one mole of quaternary compound, shown by the equation given for the reaction to be one-half of the molecular weight of potassium dichromate (147.1). The result is the number of moles of quaternary salt in the sample. Multiplying this by the empirical molecular weight (400 for the triethyl compound) gives the grams of quaternary ammonium salt in the sample. The entire calculation is thus

Percent Q =

$$\frac{(\text{Blk} - \text{titr}) \times N\,S_2O_3^{--}(\text{eq}/l) \times (\text{eq wt } K_2Cr_2O_7)(\text{g/eq}) \times \text{MWQ (g/mol)} \times 100}{\text{Sample Wt (g)} \times 1000\ (\text{ml}/l) \times \text{g } K_2Cr_2O_7/\text{mol Q}}$$

or

$$\text{Percent Q} = \frac{(\text{Blk} - \text{titr}) \times N\,S_2O_3^{--} \times 49.03 \times 400 \times 100}{\text{Sample Wt} \times 1000 \times 147.1}$$

PROCEDURE

1. Weigh a 1.5-g sample of cationic surfactant formulation, transfer it to a 100-ml volumetric flask, and dilute to the mark with water.
2. With a pipette transfer a 20-ml aliquot to a 250-ml glass-stoppered Erlenmeyer flask.
3. Transfer a 20-ml portion of standard 0.1 N $K_2Cr_2O_7$ to the flask, stopper it, and shake it for 5 min.
4. Filter the precipitate on Whatman No. 41 paper, wash it thoroughly with water, receiving the filtrate and washings in a 500-ml Erlenmeyer flask.
5. Dissolve 3 g of KI in 50 ml of water, add to it 3 ml of 6N HCl, and immediately add this solution to the contents of the Erlenmeyer flask. Swirl, cover with a watch glass, and allow it to stand in the dark for 5 min.
6. Dilute to about 400 ml, and titrate with standard 0.1N $Na_2S_2O_3$ to the endpoint of starch.
7. To determine the volume of standard thiosulfate equivalent to the dichromate, make a blank titration with the same amount of dichromate used for the sample, and calculate the percentage of cationic salt.

$$\text{Percent CS} = \frac{(\text{Blk} - \text{titr}) \times N\,S_2O_3^{--} \times 1.666 \times \text{MWCS}}{\text{Sample Wt} \times 10}$$

The calculation takes account of the 20-ml aliquot.

References

(1) Brown, P., Kitchen, J. E., and Harmston, P. 1972. Methylene bisthiocyanate: its determination and some observed effects on laboratory-scale biological sewage treatment units. *Water Pollution Control* 71:555.

(2) Few, A. V. and Ottewill, R. H. 1956. A spectrographic method for the determination of cationic detergents. *J. Coll. Sci.* 11:34.

(3) Cooling Tower Institute, Jan. 1963. *Wood maintenance for water-cooling towers.* CTI Bulletin WMS-104, p. 15.

(4) Goto, Shinko, Kawahara, Tetsuki, and Sato, Rokuro. 1963. Micro-determination of pentachlorophenol in surface waters. *Bull. Agr. Chem. Inspection Sta.*, Japan 6:19.

(5) Lowry, O. H., Rosebrough, N. J., Farr, A. L., and Randall, R. J. 1951. Protein measurement with the Folin phenol reagent. *J. Biol. Chem.* 193:265.

(6) Hill, E. C., Davies, I., Pritchard, J. A. V., and Byron, D. 1967. The estimation of microorganisms in petroleum products. *J. Inst. Petrol.* 53:275.

(7) Folin, O. and Ciocalteu, V. 1927. Tyrosine and tryptophan determinations in proteins. *J. Biol. Chem.* 73:627.

(8) Kurzer, F. 1956. Biuret and related compounds. *Chem. Rev.* 56:95.

(9) Hager, O. B., Young, E. M., Flanagan, T. L., and Walker, H. B. 1947. Determination of high molecular weight quaternary ammonium compounds as the triiodides. *Anal. Chem.* 19:885.

(10) Flotow, E. 1942. *Pharm. Zentralhalle* 83:181.

(11) Darragh, J. L. and Stayner, R. D. 1954. Quaternary ammonium compounds from dodecylbenzene. *Ind. Eng. Chem.* 46:254.

Appendix I.

GLOSSARY OF TERMS USED IN MICROBIOLOGY

Adaptation. A nonhereditary change enabling an organism to grow in an altered environment.

Aerobic. Requiring the presence of free oxygen for respiration.

Agar. A gelatinous polysaccharide, extracted from Ceylon moss, that is used to prepare solid media for cultivating bacteria.

Algae. A group of simple aquatic plants capable of photosynthesis.

Amino acid. One of a number of carboxylic acids that also contain an amino group (NH_2-).

Anaerobic. Requiring the absence of free oxygen for growth.

Anode. The positive electrode of an electrolytic cell where electrons are donated and oxidation occurs.

Autochthonous. Native to a certain habitat, such as the soil, a pond, sewage, and so on.

Autoclave. A chamber for sterilizing with steam under pressure.

Autolysis. The decomposition of a dead microorganism by its own enzymes.

Autotrophic nutrition. The process by which an organism manufactures its own food from inorganic compounds.

Bactericide. An agent capable of killing bacteria.

Bacteriostat. An agent that prevents bacterial growth by interfering with reproductive and metabolic processes.

Bacterium. One-celled, microscopic, plant-like organisms, most of which multiply by binary fission and lack chlorophyll.

Benthic. Referring to life at the bottom of a body of water.

Binary fission. Asexual reproduction by division into two equivalent parts.

Biocide. A toxicant that is fatal to living organisms.

Capsule. A mucilaginous outer layer surrounding most bacteria.

Cathode. The negative electrode of an electrolytic cell where electrons are accepted and reduction occurs.

Cell. The fundamental unit of structure in plants and animals.

Cell wall. The rigid semipermeable membrane that surrounds a microorganism, maintaining its shape and serving as a physical barrier against the environment.

Cellulose. The carbohydrate that forms the walls of plant cells.

Chain. Four or more bacteria attached end-to-end.

Chelate. A five- or six-membered ring in which a central metallic ion is bound in a coordination complex.

Chlorophyll. The green pigment in plants that catalyzes the process of photosynthesis.

Chloroplasts. Cells that contain chlorophyll and carotenes that are essential for photosynthesis.

Chromosomes. Bodies in the nucleus of a cell that bear the genes.

Ciliates. Microorganisms covered with short, whip-like extensions called cilia.

Coenocyte. An organism composed of multinucleate cells.

Colloid. A phase so highly dispersed that surface forces are an important factor in determining its properties.

Colony. A group of bacteria in a culture derived from the multiplication of a single cell; usually visible to the unaided eye.

Confluent growth. The growing together or overlapping of two or more bacterial colonies.

Conidiophore. A specialized structure in fungi that produces spores.

Counting chamber. A microscope slide with a ruled grid used for counting bacteria.

Cover glass. A thin glass slip used to cover an object mounted on a glass microscope slide.

Culture. A growth of microorganisms.

Culture medium. A preparation containing nutrients and growth factors suitable for the cultivation of microorganisms.

Cytochromes. Pigmented cellular porphyrins that transfer electrons in the process of bacterial respiration.

Cytoplasm. A colloidal suspension of proteins, lipids, and carbohydrates within a cell.

Cytoplasmic membrane. The osmotic barrier enclosing the cytoplasm serving as a controller of the transfer of metabolites and wastes into and out of a cell.

Denitrification. The reduction of nitrate or nitrite to elemental nitrogen and ammonia.

Deoxyribonucleic acid. The chemical found in chromosomes that maintains the genetic code of an organism.

Diatoms. Single-celled algae whose cell walls are composed of polymerized ortho-silicic acid.

Dispersant. An agent that lifts particles of sediment and distributes them uniformly in flowing water.

Ecology. The relationship of an organism to its environment.

Endogenous phase. The period of declining growth of a microorganism when stored food is being consumed and respiration is slowed.

Endospore. A thick-walled spore within a bacterial cell.

Enzyme. A soluble chemical manufactured in living cells that catalyzes a specific chemical transformation.

Extracellular. Outside of the cell wall.

Filamentous. Composed of long threads.

Fimbriae. Long fine hairs attached to the cell walls of certain fungi and Gram-negative bacteria.

Flagellum. A long, thread-like, rotating extension that serves as an organelle of locomotion in some bacteria.

Flocculant. An agent that causes the aggregation of small particles into large flocs.

Free-swimming. Not attached; capable of independent movement.

Fungicide. An agent that kills fungi.

Fungus. A simple plant containing no chlorophyll and not differentiated into roots, stems, or leaves.

Gene. An element of deoxyribonucleic acid that transmits a hereditary characteristic.

Genome. The totality of genetic information encoded in deoxyribonucleic acid. Also, one-half of a set of chromosomes, as in a germ cell.

Gram-negative. Bacteria that lose the initial stain in the Gram staining procedure and are red after the final stain.

Gram-positive. Bacteria that retain the initial stain in the Gram-staining procedure and are blue after the final stain.

Gram stain. A dyeing procedure used in the identification and classification of bacteria.

Habitat. The usual dwelling place of an organism.

Halophilic. Describes organisms that tolerate high concentrations of sodium chloride.

Heterotrophes. Bacteria that require complex organic compounds as sources of carbon and nitrogen.

Holozoic protozoa. Single-celled animals that ingest solid food into the body of the cell.

Humus. A complex organic mixture formed by the action of microorganisms on vegetation.

Hydrogen bond. A single-electron bond, largely electrostatic.

Hydrophilic. Having an affinity for water.

Hydrophobic. Lacking an affinity for water.

Hyphae. Multibranched tubes containing the cytoplasm of mycelial fungi.

Incubator. A chamber that maintains a constant temperature suitable for the growth and development of bacteria.

Iron bacteria. Bacteria that produce masses of slime containing hydrous ferric oxide when growing in iron-bearing water.

Lag phase. The initial period of very slow growth following the inoculation of a culture medium.

Log phase. The period of growth following the lag phase during which very rapid growth occurs.

Lysis. The leakage of cytoplasm from a cell.

Membrane filter. A nitrocellulose disk having uniform porosity and used for microbial filtration.

Mesophile. An organism that grows best at a moderate temperature—about 24–40 C.

Metabolism. The sum of the processes whereby energy is released and new cellular material is synthesized.

Metabolite. A product of metabolism.

Microbe. A microscopic organism.

Microbicide. A chemical toxic to microorganisms.

Microbiocide. Misspelling of microbicide, or else perhaps a tiny biocide.

Microbiology. The study of microorganisms.

Mitochondria. Organelles in plant and animal cells that contain the respiratory enzymes.

Mordant. A chemical used to fix a dye to some other substance.

Morphology. The structure and form of an organism.

Mucoids. Bacteria colonies that exhibit a moist glistening sheen.

Mutation. A genetic accident that physically or chemically changes an organism.

Mycelium. A mass of cytoplasm enclosed in a network of multibranched tubes.

Mycology. The study of fungi.

Nemathelminthes. Roundworms.

Nematode. A roundworm.

Nitrifying bacteria. Those bacteria that oxidize ammonia and ammonium salts to nitrate.

Obligate phototrophes. Microorganisms that cannot grow in the dark.

Ontogeny. The development of an individual living organism.

Optimum. Most favorable.

Organelle. A specialized part of a cell analogous to an organ in higher forms of life.

Osmotic pressure. The pressure inside a cell caused by the diffusion of water through the cytoplasmic membrane.

Oxidation. The addition of oxygen, the removal of hydrogen, or the loss of electrons.

Oxidative phophorylation. The conversion of adenosine diphosphate to adenosine triphosphate in the process of respiration.

Pasteurization. The partial sterilization of a liquid by heating at a relatively low temperature (80 C).

Pathogen. A parasitic organism that produces disease in a host.

Peritrichously flagellated. Flagella uniformly distributed over the surface of a microorganism.

Petri dish. A circular, shallow, covered dish used in the cultivation of bacterial colonies.

Photoautotrophes. Algae that survive in the presence of sunlight on nitrogen, carbon dioxide, and a few minerals.

Photosynthesis. The conversion of light quanta from the sun to chemical energy to sustain life.

Phylum. A major division of the Plant or Animal Kingdoms.

Phytoplankton. Floating plants such as seaweed and other algae.

Planktonic. Floating or swimming in water.

Plasmid. A doube-stranded ring of deoxyribonucleic acid that carries the genetic code of bacterial digestive enzymes; a miniature chromosome.

Plasmoptysis. The rupture of a cell wall caused by excessive intracellular pressure.

Polar flagellum. A flagellum confined to one end of a rod-shaped bacterium.

Protoplast. The living part of a bacterial cell.

Protozoan. A microscopic unicellular animal.

Psychrophile. An organism whose optimum growth occurs at low temperatures—about 4–10 C.

Quebec colony counter. A device for illuminating and magnifying a Petri dish to facilitate the counting of bacterial colonies.

Reduction. The removal of oxygen, the addition of hydrogen, or the addition of electrons.

Respiration. The release of energy by the transfer of electrons to an electron acceptor—oxygen in aerobic respiration.

Rhizopods. Protozoans that move by creeping.

Ribonucleic acid. A chemical structure found in the cytoplasm that controls chemical reactions in a cell.

Ribosome. A granule in the cytoplasm containing ribonucleic acid that controls the synthesis of protein.

Rotifers. Microscopic, multicelled invertebrates with cilia around their anterior end.

Saprophytes. Organisms that feed on nonliving organic matter.

Serial dilution. The successive dilution of a liquid sample. A 1:10 dilution is made by adding 1ml of the sample to 9 ml of water; a 1:100 dilution is made by adding 1 ml of the 1:10 dilution to 9 ml of water; a 1:1000 dilution is made by adding 1 ml of the 1:100 dilution to 9 ml of water, etc.

Sessile. Permanently fixed or unable to move.

Slime layer. A capsule that is diffuse and gelatinous.

Spore. A cell in a resistant envelop, capable of developing into a normal organism.

Sulfur bacteria. Bacteria that oxidize hydrogen sulfide or sulfur to sulfuric or sulfurous acid.

Surfactant. A chemical that alters surface and interfacial tension.

Symbiosis. A relationship between a microorganism and a host that is beneficial to both.

Synergism. A cooperative effort between two or more species of bacteria resulting in something the individual species could not have accomplished alone.

Taxonomy. Study of the general principles of scientific classification.

Tetrad. A packet of four cells, in the form of a tetrahedron, produced by successive division of a single cell.

Thallus. The body of an undifferentiated plant, such as a fungus.

Thermophile. An organism that grows best at high temperatures—about 45–75 C.

Transverse fission. Binary fission across the long axis of a bacterium.

Trichocyst. A small hair-like organelle that secretes a sticky substance in the form of threads.

Trichome. A filamentous structure.

Vegetative cells. Cells actively engaged in nutritional and growth functions.

Vegetative propagation. Plant-like growth.

Viable. Alive.

Vibrio. A Gram-negative, motile, comma-shaped bacterium.

Zoogloea. A diffuse, gelatinous mass formed by the confluent growth of bacterial capsules.

Zwitter ion. An ion carrying both positive and negative charges.

Zymogenous. Causing fermentation.

Appendix II.

QUESTIONNAIRE FOR OPERATORS OF COOLING SYSTEMS

Here is a set of questions to serve as a checklist of what a person responsible for the operation of a cooling system should know about the microbiological aspects of water treatment. The list is designed to induce the reader to think in a systematic way about the subject so that in his or her own mind there will be some sort of an answer to every question—even the rhetorical ones.

1. How often does your water service company representative call?
2. Do you have regular discussions with him?
3. Are these helpful to you? How?
4. Do you receive regular written reports from your service representative?
5. Are these helpful to you? How?
6. Are they accurate? How do you know?
7. Do you receive reports of microbiological analyses?
8. How do you use them?
9. How does your service representative say he uses them?
10. What is your program for microbiological control?
11. How do you evaluate the results of this program?
12. Is it effective? How do you know?
13. Do you have difficulties with microorganisms in your cooling system?
14. Do you chlorinate your cooling system?
15. Does it do any good? How do you know?
16. What biocides other than chlorine do you use?
17. How do you decide when to use them?
18. How did you choose them?
19. How can you tell whether they are effective or not?
20. What is the current monthly cost of biocides for your system?
21. Do you receive reports of analyses of scales and fouling deposits?
22. Are these of any use to you?
23. Are any changes ever made in the treatment program on the basis of

results obtained from microbiological or deposit analyses?

24. Are corrosion coupons used in your system? Do you find the results useful? How?
25. Do you have a regular schedule for the structural maintenance of cooling towers?
26. Do you have facilities for back-washing heat exchangers? Do you use them?
27. Do you check heat transfer coefficients on a regular schedule?
28. Do you keep records of chemical cleaning and hydroblasting?
29. What is your procedure in the event of an acid leak or contamination of cooling water by an acidic gas?
30. What is your procedure in the event of leaks of reducing agents such as hydrogen sulfide or sulfur dioxide?
31. What does your service representative advise when leaks occur?
32. How do you know when you have leaks?
33. Do you have facilities for obtaining chromatographic "fingerprints" to identify the sources of oil and organic chemical leaks?
34. What is the current monthly cost of treatment for your cooling system?
35. What are the current unit prices of your water-treating chemicals?
36. Do you use the same water treatment program in all of your cooling systems? If not, why not?
37. Do you consider your current cooling water treatment program satisfactory?

Appendix III.

NUMERICAL PROBLEMS

1. The killing rate constant of a certain biocide is 0.014. If the bacteria count in a sample of cooling water is 6.0×10^6/ml, what concentration of the biocide should be used to reduce the count to 1.0×10^5/ml in 6 h?

Ans. 48.7 ppm

2. If the contained volume of the cooling system in the foregoing example is 425,000 gal, how many pounds of the biocide should be used?

Ans. 173 lb

3. Two microbicides, A and B, have killing rate constants of 0.026 and 0.011, respectively. The unit price of A is $2.75/lb, and that of B is $1.16/lb. If it is desired to reduce the bacteria count in a cooling water from 3.0×10^6 to 1.0×10^5/ml in 6 h, which microbicide is the more economical to use?

Ans. Essentially the same cost.

4. A service company respresentative proposes to treat a cooling system with a polymer which, he says, will disperse silt, if added to the recirculating water at the optimum concentration of 1 ppm. The contained volume of the system is 300,000 gal, and the blowdown rate is 1875 gph. The representative suggests adding 2.5 lb of polymer daily. If this advice is followed, what will be the concentration of polymer when equilibrium has been reached?

Ans. 7.1 ppm

5. How many days will it take to reach the equilibrium concentration?

Ans. 33.9 days

6. How many pounds of polymer should really be added to keep its concentration at 1 ppm maximum?

Ans. 2.5 lb 1st day, then 0.35 lb daily.

7. In carrying out a toxicant evaluation of a certain biocide as described on pp. 200–201 the following results were obtained after 6 h of exposure: At 0 ppm the plate count was 3.2×10^6; at 5 ppm it was 1.3×10^6; at 7.5 ppm it was 830,000; at 12.5 ppm it was 330,000. What is the killing rate constant of the biocide?

Ans. 0.030

8. What concentration of the above biocide should be used to reduce the bacteria from 3.2×10^6 to 1.0×10^5/ml in a period of 6 h?

Ans. 19.3 ppm

9. If a plate count of 3.2×10^6 were obtained in a cooling system containing 425,000 gal of water, what weight of the biocide should be added to reduce the count to 1.0×10^5/ml?

Ans. 68.3 lb

10. If the density of the biocide solution evaluated in problem 7 is 9.75 lb/gal, how many gallons should be added to give a concentration of 19.3 ppm in the 425,000-gal cooling system?

Ans. 7 gal

Index

www.ingramcontent.com/pod-product-compliance
Lightning Source LLC
LaVergne TN
LVHW101442200926
848483LV00001B/26